Socially Just Mining: Rethoric or Reality? Lessons from Peru

Synthesis Lectures on Engineering, Technology and Society

Editor
Caroline Baillie, *University of San Diego*

The mission of this lecture series is to foster an understanding for engineers and scientists on the inclusive nature of their profession. The creation and proliferation of technologies needs to be inclusive as it has effects on all of humankind, regardless of national boundaries, socio-economic status, gender, race and ethnicity, or creed. The lectures will combine expertise in sociology, political economics, philosophy of science, history, engineering, engineering education, participatory research, development studies, sustainability, psychotherapy, policy studies, and epistemology. The lectures will be relevant to all engineers practicing in all parts of the world. Although written for practicing engineers and human resource trainers, it is expected that engineering, science and social science faculty in universities will find these publications an invaluable resource for students in the classroom and for further research. The goal of the series is to provide a platform for the publication of important and sometimes controversial lectures which will encourage discussion, reflection and further understanding.

The series editor will invite authors and encourage experts to recommend authors to write on a wide array of topics, focusing on the cause and effect relationships between engineers and technology, technologies and society and of society on technology and engineers. Topics will include, but are not limited to the following general areas; History of Engineering, Politics and the Engineer, Economics , Social Issues and Ethics, Women in Engineering, Creativity and Innovation, Knowledge Networks, Styles of Organization, Environmental Issues, Appropriate Technology

Socially Just Mining: Rethoric or Reality? Lessons from Peru
Caroline Baillie, Eric Feinblatt, Joel Alejandro Mejia, Glevys Rondon, Jordan Aitken, Rita Armstrong, Vicki Bilro, Andy Fourie, and Kylie Macpherson
2020

Drones for Good: How to Bring Sociotechnical Thinking into the Classroom
Gordon D. Hoople and Austin Choi-Fitzpatrick
2020

Engineering Ethics: Peace, Justice, and the Earth, Second Edition
George D. Catalano
2014

Engineering and Society: Working Towards Social Justice, Part II: Decisions in the 21st Century
George Catalano and Caroline Baillie
2009

Engineering and Society: Working Towards Social Justice, Part III: Windows on Society
Caroline Baillie and George Catalano
2009

Engineering: Women and Leadership
Corri Zoli, Shobha Bhatia, Valerie Davidson, and Kelly Rusch
2008

Bridging the Gap Between Engineering and the Global World: A Case Study of the Coconut (Coir) Fiber Industry in Kerala, India
Shobha K. Bhatia and Jennifer L. Smith
2008

Engineering and Social Justice
Donna Riley
2008

Engineering, Poverty, and the Earth
George D. Catalano
2007

Engineers within a Local and Global Society
Caroline Baillie
2006

Globalization, Engineering, and Creativity
John Reader
2006

Engineering Ethics: Peace, Justice, and the Earth
George D. Catalano
2006

Socially Just Mining: Rethoric or Reality? Lessons from Peru

Caroline Baillie, Eric Feinblatt, Joel Alejandro Mejia, Glevys Rondon, Jordan Aitken, Rita Armstrong, Vicki Bilro, Andy Fourie, and Kylie Macpherson

ISBN: 978-3-031-00989-1   paperback
ISBN: 978-3-031-02117-6   ebook
ISBN: 978-3-031-00168-0   hardcover

DOI 10.1007/978-3-031-02117-6

A Publication in the Morgan & Claypool Publishers series
SYNTHESIS LECTURES ON ENGINEERING, TECHNOLOGY AND SOCIETY

Lecture #25
Series Editor: Caroline Baillie, *University of San Diego*
Series ISSN
Print 1933-3633   Electronic 1933-3641

Socially Just Mining: Rethoric or Reality? Lessons from Peru

EDITORS and AUTHORS

Caroline Baillie
University of San Diego

Eric Feinblatt
Director of Waste for Life

Joel Alejandro Mejia
University of San Diego

Glevys Rondon
Independent Researcher and Consultant

AUTHORS

Jordan Aitken, Rita Armstrong, Vicki Bilro, Andy Fourie, and Kylie Macpherson

SYNTHESIS LECTURES ON ENGINEERING, TECHNOLOGY AND SOCIETY #25

ABSTRACT

In this book we consider ways in which mining companies do and can/should respect the human rights of communities affected by mining operations. We examine what "can and should" means and to whom, in a variety of mostly Peruvian contexts, and how engineers engage in "normative" practices that may interfere with the communities' best interests. We hope to raise awareness of the complexity of issues at stake and begin the necessary process of critique—of self and of the industry in which an engineer chooses to work. This book aims to alert engineering students to the price paid not only by vulnerable communities but also by the natural environment when mining companies engage in irresponsible and, often, illegal mining practices. If mining is to be in our future, and if we are to *have a future* which is sustainable, engineering students must learn to mine and support mining, in new ways—ways which are fairer, more equitable, and cleaner than today.

KEYWORDS

mining and social justice, human rights defenders, gender rights, gender violence, environmental justice, Peru, mining, progress, development, conflict, water, contamination, tailings, environment, land rights, FPIC, corporate social responsibility, community

Contents

3 The Ineffectiveness of Human Rights Protection Mechanisms for Communities Affected by Mining: A Case Study of Minas Conga in Cajamarca, Peru

Jordan Aitken

4 Exploring the Notion of Socially Just Mining Through the Experiences of Five Indigenous Women from Latin America

Kylie Macpherson

7   Translating Values into Action: What Can Be Done? **173**

Rita Armstrong, Caroline Baillie, Andy Fourie, and Glevys Rondon

Acknowledgments

Our team would like to thank the IM4DC (International Mining for Development Centre) at the University of Western Australia for funding the trips to Peru that made much of the work possible. We would also like to thank everyone who shared their time and views with us as they see this as such an important area of work.

Caroline Baillie, Eric Feinblatt, Joel Alejandro Mejia, Glevys Rondon, Jordan Aitken, Rita Armstrong, Vicki Bilro, Andy Fourie, and Kylie Macpherson
September 2020

CHAPTER 1

Introduction

Engineers are passionate about technology, how to use it and how to develop it. As engineering students, we learn about mining as a continuous process of extraction. We also learn about how ore is transformed through various processes before it is mined and made into a metal. As mining and petroleum engineers we learn about the technical details of the extraction process. When analyzing specific minerals, we investigate how to extract the ore so that it is financially viable while removing "contaminants" that make the metal "impure." As mechanical, chemical, materials, civil, and electrical engineers, we might be interested in the processing of these minerals and oils, or we might just use them to make our smart materials, our bridges, our cars, and our soaps. We learn about health and safety issues and risk assessment of our models in general, but we don't consider the real impact on specific groups of people. We learn about the importance of water and the role it plays in all mining operations—we learn about grinding, flotation, cyclones, separation, electrowinning, and concentration but we don't learn about how people are affected by the depletion of water. In Jujuy, Argentina everyone in mining circles talks about the big lithium deposits, essential for the production of smartphones and electric car batteries. But the Kolla people have challenged lithium exploitation not only because they have not been consulted but also because they fear water shortages in this very arid area. Some may learn about ethics and environmental impact but in generic ways with no real appreciation of the details of how peoples, their lands, and ecologies are disrupted. Few engineering students are exposed to the day-to-day pain that people suffer who live in close proximity to a mine site. As Bruce Harvey formerly Global Practice Leader—Communities and Social Performance at Rio Tinto, told us:

> *...The irony of course is that we would never dream of developing a mineral resource without getting a comprehensive picture of the resource, drilling it out at a requisite drill spacing, understanding the chemical composition, understanding the material handling characteristics of the ore itself, we would never dream of building a large coal washery without actually doing a lot of very expensive bench testing on the wash characteristics of the coal. Yet regularly we go into places and spend billions of dollars without understanding the social fabric in which we're working in which we now know is the most important fabric right?* (Harvey, 2014[1]) [1].

[1]Where a quote is attributed to a person and date with no associated reference, it refers to interviews conducted as part of our research studies in Peru during the years 2014–2016 as discussed in Chapters 2 and 7, with permission given by interviewees to include their names.

Although mining is sometimes understood—or even misunderstood—by many and perceived as separate compartmentalized operations that are isolated from social, cultural, historical, or economic contexts, mining companies know full well the financial consequences of not engaging properly with local communities. This knowledge has had a profound impact on the industry. Tony Hodge, former President of the International Council for Mining and Minerals described to us a recent shift in requirements of mining companies for their employees:

> *There has been a shift in values inside mining companies now… what the senior executives of mining companies are really looking for in young engineers for example that has changed dramatically, hugely. You can get design specialists, you can get computer wonks, you can get all of that, what they really need are people that understand all of that and know how to engage effectively and build relationships and that's a tough sell, tough ask …what our mining companies need most now are people that know how to build relationships. But you have to do that with the knowledge, the mining process. You have to have as much as fundamental understanding of electrical engineering or civil engineering or chemical engineering or whatever it happens to be, geological engineering, but you have to be able to respect people, you have to understand how to interact effectively with people because companies are working in different cultures. …You've got to have the skill at it, the sensitivity of it. Have to learn just not to listen but hear and those are skills that are not typical in our engineering faculties* (Hodge, 2014) [2].

But Bruce Harvey is quite clear here—we don't need to become a social scientist to have this new skill, but we do need to be aware enough of the issues to know when we need expert advice and understand this advice when it is given. We do need to start including these elements in school:

> *Step one, acquire a comprehensively verified validated understanding of the social landscape in which we're developing. Then step two is form partnerships with people who know how to do the work that we don't know how to do…There's another whole set of data that's more qualitative, and we're talking the data that anthropologists would collect you know—how do people live their lives around here? Where does power come from? How are decisions made? How are the checks and balances in civil society around here undertaken? What role do cultural or religious authorities have vs. civic authorities? What are the customary norms that prevail?…understanding the norms of people's lives in households and in extended families is a very important part of understanding how we will achieve societal stability in order for us to run a multigenerational mine without it becoming a victim of chaos or anarchy or civil discontent* (Harvey, 2014) [1].

Harvey believes that if engineering students were able to think a bit more broadly, that they would be able to bring in their excellent "disciplined thinking" and apply this to a wider range of social issues:

I mean they're like fighter pilots, they can focus on something for a very, very long period of time and be very vectored in the way that they think and that's a great attribute, and if we were able to bring that and use it into other areas which are increasingly important in the mining industry, we would have solutions that we wouldn't have otherwise thought about (Harvey, 2014) [1].

This whole book series is dedicated to developing the awareness, knowledge, and skills of students in the interface between society, environment, and engineering. If Harvey thinks this should be done, why is it not being done? Being able to develop a critical consciousness, to question "traditional" assumptions about what an engineer does in society and to develop empathic skills and an awareness of the huge impact that engineering has on peoples' lives, is key to developing these capabilities. Harvey describes his own background education in this area as key to his success:

When I did my geology degree 35 years ago, I'm not even quite sure how it happened but thank goodness it did. I did a unit and through to the second year in history and philosophy including the history and philosophy of science and it was really fun. I mean there were only four of us in the science stream who were in the class and the rest were all humanity students, but the things that we discovered and the way that we learned to understand how the philosophy of science and science and now natural science evolved over the last 2,000 to 3,000 years but most particularly during the enlightenment, it's affected everything that I did subsequently you know and as a apart from this outcome scientist I'm constantly exposing myself to personal and peer review. The answers we come up with I know are not the truth, you know more often than not they're a satisfying approximation of the truth and they work extremely well whereas I think a lot of my engineering and science colleagues who didn't expose themselves to philosophy, and particularly unpick the philosophy of science as expressed by and analyzed by people like Popper and Kuhn and others, they've got an overwhelming almost arrogant presumption in the answers that they generate through mathematics and that leads to many of the mistakes that we experience you know in project design and project implementation because there's not enough skepticism about the results that are generated. There's not enough exploring of the various scenarios and options that could work because we're so convinced that this linear method of engineering design can generate one result and it's only that result which is right. So the ability to self-critique and self analyze and understanding the fallibilities and the falsify ability a doctrine that lies behind science, is something that I think many of our engineers could really benefit from (Harvey, 2014) [1].

Students need to develop the capabilities of dealing with real issues happening out there where there is no numerical answer. They need to develop critical consciousness or develop empathic skills to enhance their professional practice. Enabling students to be able to understand better the social fabric—and listen to not only the perspectives of the mining companies and government but also the local communities would be a game changer. In this book we bring you some of the views of these stakeholders in the mining sector in Peru. We demonstrate that actors in all these spaces claim to want the same thing: peaceful community engagement with no conflict. They also make similar claims about how this might occur. However, as expressed by Tony Hodge, with the best will in the world, company CEOs may not in fact be in control of how this engagement is enacted:

> *In fact most people don't understand the mining industry. As best we can estimate, there's 21 companies on our council, there are about 6,000 mining companies in the world…our 21 companies employ about a million people out of two point five million in the formal sector, that doesn't include the miners who can range in numbers from 30 to a hundred million depending on what's going on in the world and nobody really knows… So you've got a little corporate office and you've got operations all over the world and …it's probably the greatest challenge to bringing a lot of these ideas into real practice than any, any other issues that arise. Not that we don't have the intellectual concepts pinned down or that people don't have the recipes for doing it differently, these are changes in practice that come through introducing change inside companies that are very, very large companies and they're internal administrations suffer from all of the strengths and weaknesses of very large agglomeration of people* (Hodge, 2014) [2].

If operations get so large that even those companies who claim to wish to do the right thing are not in control of how their own staff engage with local communities, and in countries where governments do not protect the rights of their own citizens, conflicts will and do arise.

In this book we consider ways in which mining companies can and should behave, even when the going gets tough, and even when national and local governments fail in their duty to protect, so that they can effectively respect the human rights of communities affected by their mining operations. We will examine what "can and should" means and to whom, and how engineers engage in "normative" practices that may interfere with the communities best interest. We hope to raise awareness of the complexity of issues at stake and begin the necessary process of critique that Harvey suggests—of self and of the industry in which a student chooses to work. This is the second book within the Engineers, Technology, and Society series on this topic. The first, *Mining and Communities: Understanding the Context of Engineering Practice*, focused on mine sites in Papua New Guinea and Australia. This second book explores a very different context within Latin America, focusing mostly on Peru. While the country, traditions, and history frame a different community's response to mining challenges, we find that much of the underlying concerns on people and the environment remain very similar.

We begin by sharing stories from communities in Peru that have experienced environmental disasters on their land (Chapter 2). We look at the current state of play regarding the right of communities to exercise their fundamental rights, such as the rights to a safe environment, to water and self-determination, and how, as a result of their importance, the Peruvian constitution grants its citizens the right to defend and protect themselves through peaceful protests. What we show here (Chapter 3) is that often communities demand that the recognition of these rights should mean that they are able to establish their own development outcomes, and that projects perceived as damaging to the environment, their livelihood, and, ultimately, their survival should be rejected by the government. The challenge is that local communities expect their decisions to be accepted by corporations as well as part of Free, Prior, and Informed Consent (FPIC), which is recognized under international law as best practice for development projects. What we see among the different cases reviewed in this book is that communities are using key legal tools available to them to transform consent into a right, and it is in this context that protests are intimately connected with power and consent.

Chapters 4 and 5 introduce us to the price that local activists pay for reclaiming their rights, defending their land and lives. We also explore the obstacles communities face when communities try to have access to remedy and reparation for wrongdoings made by mining companies (Chapter 6).

In our final Chapter 7, we take a deeper dive into the issues in Peruvian mine sites and ask the views of mining companies, non-governmental organizations (NGOs), community-based organizations, farmers, and government representatives what can and should be done to stop the abuses. We finish with our own set of guiding principles, some short films and a Mining and Communities game, which can be used in educational or human resource training contexts to help raise awareness of a multitude of contextual mining issues to emerging engineers.

The whole book aims to alert engineering students to the price paid not only by vulnerable communities but also by the natural environment when mining companies engage in irresponsible and, often, illegal mining practices. If mining is to be in our future, and if we are to *have a future* which is sustainable, engineering students must learn to mine and support mining, in new ways. Ways which are fairer, more equitable, and cleaner than today.

1.1   REFERENCES

[1] Harvey, B. (2014). Social development will not deliver social licence to operate for the extractive sector. *The Extractive Industries and Society*, 1:7–11. DOI: 10.1016/j.exis.2013.11.001. 1, 2, 3

[2] Hodge, R. A. (2014). Mining company performance and community conflict: Moving beyond a seeming paradox. *Journal of Cleaner Production*, 84:27–33. DOI: 10.1016/j.jclepro.2014.09.007. 2, 4

CHAPTER 2

Mines and Community Engagement in Peru: Communities Telling Their Stories to Improve Future Practice

Rita Armstrong, Caroline Baillie, Andy Fourie, and Glevys Rondon

2.1 INTRODUCTION

The surge in mining activity in Peru since the 1990s has been accompanied by escalating social conflict which, despite state and corporate emphasis on social responsibility and sustainability, has not abated. This chapter examines the community experience of mining in Peru with the aim of understanding the underlying issues which have led to a breakdown in community engagement, and to identify community recommendations for improvements in that area. The ultimate goal is to raise awareness of the need for an improvement in community-mine relations given the current high level of social protest, conflict, and violence.

Two mine sites are compared in this study: one with a steady escalation of social conflict (Yanacocha) and one which despite utilizing third-party non-governmental organization (NGO) mediation and a roundtable dialogue, also generated social conflict (Tintaya). Interviews were conducted with a selection of community members at both mine sites. The main purpose of the interviews was to understand whether they felt "engaged," respected, and "heard" in their engagements with miningcompanies. Each interview was open-ended but guided by a common structure and guiding questions which are set out in Appendix A. All people who agreed to be interviewed[1] expressed themselves with a degree of honesty and transparency that would not have been possible without the unique capability of the researcher, Glevys Rondon, who has been visiting both regions for the past decade. All interviews were conducted in Spanish.

---

[1]Ethics approval for the above project was granted by The University of Western Australia: # RA/4/1/6548.

A brief history of mining in Peru will provide the background for a more detailed account of the social and economic conditions in each mine site: Yanacocha and Tintaya. Detailed narratives are provided by men and women who have been impacted by mining at each site which provide a personal account of the first-hand experience of different aspects of the mining process from optimistic expectations to unease and, in many cases, anger. The final section of the chapter compares the shared experiences of these two communities, many of which resonate with mining histories in other parts of South America, and makes recommendations for guiding principles which must shape relations between mines and communities.[2]

2.2 MINING IN PERU

2.2.1 GOVERNMENT POLICIES TOWARD MINING

There is a long history of mineral extraction in Peru associated most famously with the Inca empire and then with Spanish colonization in the middle of the 16th century. The Spanish conscripted local Peruvian labor (in the *mita* system) to work on the mercury mines at Huancavelica and on the Potosi silver mines in Bolivia (Dell 2010) [22]. Pizarro, the leading conquistador, has gained notoriety for his treatment of the Inca emperor Atahualpa at Cajamarca when he announced that Atahualpa would regain his freedom if he had a chamber filled with of gold for the Spaniards. Once Atahualpa had acceded to that demand, Pizarro murdered him. The influx of European companies in the Andes and effectively enslaved those who lived there—with consequences which still vitally affect Peruvian society and nurture deeply rooted resentments (World Bank 2005, p. 17) [53]. Lima became the administrative hub of the Spanish empire and the Spanish created a legacy of "an authoritarian tradition of government led by conservative local elites and largely impervious to social pressures from below" (Crabtree and Crabtree-Condor 2012, p. 49) [17]. The electoral reform law in 1896, for example, excluded anyone who could not read or write from voting—essentially most of the rural and Indigenous communities—and this law was not amended until 1979.

The International Council of Mining and Metals identifies three distinct periods in the Peruvian mining sector (ICMM 2007) [26].

- Prior to the late 1960s, mines were largely privately owned. Many U.S. international corporations controlled large-scale mining while local entrepreneurs concentrated on small to medium mining projects (Echave 2005, p. 117) [20].

- After 1968, the military regime established state control of the mining sector, which continued until the 1980s and the return of democratic government.

- The social upheaval caused by the Shining Path insurgency movement, accompanied by high inflation, facilitated a return to more authoritarian yet laissez-faire government

---

[2]This material is based on action research carried out for the IM4DC (International Mining for Development Center at the University of Western Australia and the University of Queensland) and is largely derived from the final report.

under Fujimoro. Fujimoro implemented a series of what have been described as neoliberal measures: eliminating price controls, government subsidies, and offering legal and financial protections for foreign investors (Bury 2004, p. 80) [13]. Investment in exploration increased by 20% in Peru between 1990 and 1997 and by 2003 "mining accounted for 57% of all exports in Peru, and 37% of foreign direct investment between 2001 and 2003" (Bebbington et al., 2007, p. iv) [7].

2.2.2   MINING AND RURAL SOCIETY IN THE ANDES

By 2010, over half of Peru's *campesino* (peasant) communities lived in mining affected area (Bebbington et al., 2007, p. iv) [7]. Rural peasants in the Andes have historically relied on a broad range of activities (including pastoralism and horticulture) to sustain their income. Besides working in livestock production they also engage in petty trading and migrant wage labor (Bury 2007) [15]. In this region, access to water and communal decision-making about land use is a crucial determinant of people's ability to make a living in markedly different altitude zones (Braaten 2014) [11].

The Peruvian military government instituted agrarian land reforms in the mid-1970s which redistributed land from the haciendas according to the principle that "land belongs to those who work it" but land reforms did not necessarily ease the hardship experienced by the *campesinos* (Kay 2007). In terms of organization, the *campesinos* worked individually on separate land plots, while functioning as local communities. Water management is one area which requires community co-operation and collective action (Braaten 2014, p. 129) [11]. It cannot be overemphasized that throughout centuries of dispossession, the Andean people have continued to use a subsistence system which has enabled them to "confront the risks posed by an agriculture weighed down by the limitations of the Andean environment" (Cusicanqui 1993, p. 79) [18].

2.2.3   MINING AND SOCIAL CONFLICT

In conventional economic terms, the mining sector has increased the revenue of the Peruvian government, led to greater economic stability, and decreased inflation (ICMM 2007 [26], Bebbington et al., 2007, p. 3 [8]). There has not, however, been a commensurate improvement in local economies or living conditions: this type of disparity is not uncommon and is often referred to as the "resource curse." What is more uncommon and more disturbing, is that this recent surge in mining activity has also been accompanied by a steady rise in social conflict which, despite state and corporate emphasis on social responsibility and sustainability, shows little signs of abating.

The following two sections of report will outline two cases of company-community conflict: at Yanacocha and Minas Conga in the northeast and at Tintaya in the southern Andes.

2.3 YANACOCHA MINE

2.3.1 THE CAMPESINOS OF CAJAMARCA

In 1993, one year after construction of the mine began, more than 96% of the Department of Cajamarca were rural peasants or *campesinos* (Bury 2004, p. 81) [13]. Bury, a human geographer, describes current conditions as one of "extreme poverty" although his measures of poverty are based on categories which bear little or no relation to the indigenous categories of lack or deprivation. We are told for example that per capita income is less than half of the national average and that most of the houses do not have water or electricity, and that roads are almost non-existent. We are also told that 80% of the population cannot meet basic needs but we do not know how these needs are defined. He goes onto cite authors Gonzales and Trivelli who claim that the people of Cajamarca live "in a state of misery" (1999, p. 97) [13]. Bury's description of rural life does, however, convey a sense of hardship and this is due to many historical factors some of which set out below.

Cajamarca was the focal point of Spanish colonization, led by Franciso Pizarro who infamously captured the Incan ruler Atahualpa, and subsequently betrayed him. Under colonial rule, Cajamarca became a major textile center, then a mining center with the discovery of silver in Hualgayoc in 1772. The region was then developed into a hacienda system of feudal class relations which expanded at the expense of the land base of Indigenous communities (Deere 1990, p. 24) [21]. With the Spanish colonization of Latin America, lower classes of Spanish immigrants "could 'buy' their way into aristocracy as soon as they had accumulated enough money to pay for it." By 1940 Cajamarca had the highest concentration of peasants living under the *hacienda* system: 46 haciendas owned 65% of the land (Deere 1990, p. 27) [21].

After the Colonial period, once the Spaniards were expelled from the Andes, the land did not return to the Indigenous population but stayed in the hands of creoles and mestizos who perpetuated the hacienda system. In the Cajamarca region, this history of rural and political disenfranchisement has laid the foundation for collective action, in the form of peasant associations: the *rondas campesinas*.

The rondas have been described as an "exceptional phenomena," formed in response to "feelings of disappointment and distrust concerning the official system of justice" (Munoz et al., 2007, pp. 1932) [42]:

> "Furthermore, campesinos (peasants) allege that they are treated with contempt. They have to wait till last to be seen by public officials, they are tricked because they cannot read or write, and they have to show deference for those titled 'doctor,' 'boss,' or 'sir'" (Munoz et al. 2007, pp. 1932–33) [42].

As we shall see from the narratives below, these experiences and feelings still inform *campesina* perceptions of current governments and mining companies.

2.3.2   THE YANACOCHA AND MINAS CONGA MINE SITES

The Yanacocha mine, operated by Minera Yanacoca S.A. (MYSA), is the largest open pit gold mine in Latin America. MYSA is a joint venture comprising Newmont Mining Corporation (51.35%); Condesa, a subsidiary of the Peruvian company Minas Buenaventura (43.65%); and the International Finance Corporation or IFC (5%). Newmont began construction of Yanacocha in 1992. It was the first large foreign investment in Peru since 1976 (Bury p. 80) and by 2000 MYSA was Latin America's largest gold producer. At that time, the mining sector accounted for 45% of all national exports (IMF 2001, Bury, p. 80) [42].

Minas Conga is an expansion of Yanacocha. The project is part of the development of a larger mining district that contains different copper and gold deposits, most of which belong to MYSRL. The project area straddles the Sorochuco and Huasmín districts of the Celendín province (also part of the Cajamarca Region) and the district of La Encañada in the Cajamarca province (Kemp et al. 2013, p. 1) [30].

Many residents of local communities are opposed to the project on the grounds that it will destroy multiple high Andean lakes and threaten their access to sufficient, safe, and affordable water, on which they depend for farming, livestock, and human consumption. Moreover, they claim the right to determine their own regional development and argue their right to free prior and informed consent has not been respected.

The following timeline sets out the escalating conflict between MYSA and the communities impacted by mining activities. This historical sequence will provide a contextual backdrop for the interview material in Table 2.1.

This timeline reveals early dissatisfaction about the impact of the Yanacocha mine on water sources and water quality which was exacerbated by the mercury spill, followed by anger at plans to develop another mine stite at Minas Conga. When that plan is approved in 2010, conflict escalates into massive protests and the declaration of a State of Emergency in 2012. An outsider who reviewed this timeline might be puzzled about this chain of events. After all, Newmont instigated a "citizenship participation process" in 2007 with regular "Days of Dialogue" for residents to come to the Public Information Office and have their opinions heard and recorded. This process apparently satisfied the Ministry of Energy and Mining which approved the EIA for the proposed mine site at Minas Conga.

So we have to ask ourselves: what kind of community engagement took place before 2007, why didn't the Days of Dialogue alleviate community concerns, and what kinds of emotions drove local people to take to the streets of Celendin in the face of massive police presence? The narratives below will provide some answers to these questions by revealing personal experience of a range of issues: the early days of mining, of the "Days of Dialogue," and of State response to social protest.

Table 2.1: Timeline for MYSA escalation of conflicts in Peru (*Continues.*)

Date	Event
1993	Mining commences at Yanacocha, to the northeast of Cajamarca.
Late 1990s	Concerns about quality of urban water supply among Cajamarca residents.
Late 1990s	Exploration lease for Minas Conga project and purchase of land for that project by MYSA.
Late 1990s	Peasant protests against exploration of area known as Cerro Quilish.[*]
June, 2000	Accidental mercury spill in the transportation of chemicals to Yanacocha running into the streets of a nearby village and some of the residents collected it, thinking it had some economic value, thus exposing themselves to serious health risks. MYSA reported that it had recuperated 147 of the 150 kg that had been spilled.
2004	Constitutional Tribunal allows MYSA to recommence exploration at Cerro Quilish and protests erupt in Cajamara; MYSA withdraws its application.
2007	The Conga Project Team established the Citizen Participation Process ("Proceso de Participacion Ciudadana," or PPC) to ensure constructive dialogue while developing the Conga Project's EIA. With a goal of "transparency and inclusiveness," Conga's PPC offered individuals and community groups a variety of mechanisms in which to participate in the process, including: (1) Public Information Offices—established to offer a central location for community members to communicate with the Conga Project Team and discuss project-specific issues, and (2) Days of Dialogue—scheduled dates and times at the Public Information Offices for citizens to meet with Project team members and have their comments and ideas recorded as part of the Conga EIA review process.
2010	EIA approved for Minas Conga "following a three-year, public participation process on the project's Environmental Impact Assessment (EIA) and extensive reviews by 12 government agencies."
May, 2011	Jaime Chaupe from Sorochuco, filed a claim against MYSA for invading his family's land.
June, 2011	Ollanta Humala becomes President promising more social inclusion. Oxfam reports that Humala was elected "with a fair amount of hope that he could provide a solution to these conflicts, but much remains the same."
Sep., 7 2011	President Ollanta Humala approves Indigenous Consultation Law.

[*] Cerro Quilish "is a small mountain that comprises the top of the watershed supplying the city of cajamarca and the valleys of the Porcon and Grande rivers" (Triscitti 2013, p. 439).

Table 2.1: (*Continued.*) Timeline for MYSA escalation of conflicts in Peru

Dec. 5, 2011	President Ollanta Humala declares State of Emergency in Cajamarca. He blamed the impasse over the project on local officials: "Every possible means has been exhausted to establish dialogue and resolve the conflict democratically, but the intransigence of local and regional leaders has been exposed—not even the most basic agreements could be reached to ensure social peace and the re-establishment of public services," he said.
2012	In 2012, an independent panel of international experts reviewed the EIA and confirmed the project's original EIA met Peruvian and international standards. It is the Ministry of Energy and Mining (MEM) that has final say, however.
July 3, 2012	3,000 people marched in protest in Celendin against Minas Conga. In subsequent clashes between demonstrators and security forces, at least 20 civilians were shot with live ammunition, 4 of them fatally.
July 3, 2012	State of emergency declared in Celendin.
Aug., 2012	MYSA commissions the CSRM at UQ to get community feedback from residents of Cajamarca.
Aug., 2012	The Peruvian government restructured its conflict management office, and renamed it the National Office of Dialogue and Sustainability. The aim of this office is to address conflict in a broader community development context, rather than only responding to social conflicts after they have already erupted.

2.3.3   NARRATIVES ABOUT THE EXPERIENCE OF MINING IN CAJAMARCA AND CELENDIN

Those who were interviewed included the following:

1. Male resident of Cajamarca (YU)

2. Female resident of Cajamarca (YV)

3. Female resident of Celendin (YW)

4. Male resident of Celendin (YS)

5. Female *campesino* of Sorochuco (YC)

6. Male *campesino* of Celendin (YR)

2.3.4   NARRATIVE 1: MALE RESIDENT OF CAJAMARCA

YU is a male university graduate whose experience with MYSA comes from being a resident of Cajamarca since the mine began, and being involved in a local NGO group that monitors mining activities. He acknowledges that "Cajamarca received Yanacocha with a certain happiness because they were bringing new jobs" and that it was only with hindsight, seeing how the company operated and how the projects expanded to become a much bigger operation, that urban and rural residents began to form a very different opinion of MYSA. He also acknowledged that some people support mining but these are people who are not directly impacted by its activities. He identifies three key issues that catalyzed public opinion against MYSA: the low prices paid for land and the means by which they acquired it around Yanacocha; the subsequent mercury spillage at Choropampa; and the plan to mine at Cerro Quilish.

Encounter with MYSA: Initial Optimism Turns Sour

YS describes how he initially "spent a lot of working hours speaking with the people from the social responsibility, interchanged ideas, we started to think about things that could be done to improve communication." The NGO that he worked with initially had the "view that a company could eventually develop activities without violating human rights or affecting the environment so for some time we had contact with them, we participated in the forums which the company organized ...." But later, he realized that the company had "two faces." The "nice" face of the company is the community relationships section:

> "they hire women for the role of community relationships. In general for community relation they hire people with good intentions, anthropologists for example, but also many women, so those in charge of social responsibility are women in a lot of the cases ... and these social responsibility people try to do things in one way, in a coherent way, according principles that they believe in but they are not aware of all of the other dimensions of the company."

He describes community relationships personnel like "replacement parts of a machine" because the people "are rotating all the time." The real power, he says, resides in the "other face" of the company, that is, with the people who ultimately make decisions which alienate communities. His views about these decisions are summarized below.

Misinformation About Minas Conga

He agreed that MYSA instigated a dissemination programme but that "they have shown only what they wanted to show ... so it was more like propaganda." He describes the brochures about Minas Conga as "very general" and unrealistic because "they say how good the water will be treated, how good the environment will be treated, what will be gained, right?" He also felt that the Thursday Dialogue sessions were not genuine dialogues because people had to write down issues that bothered them "but they do not receive an answer and the study, basically it is approved, so it is only a formality, to show it to the community. So there is hollowness to these proceedings…."

Increased Repression: The Company and the Government in Alliance
One of the main changes he has observed since the beginning of MYSA's operations is that the company, instead of trying to change the face of community engagement—to being more conciliatory, for example—is that it has chosen a more repressive and adversarial approach. This is evident, he says, in the use of military consultants, and in refusing to recognize the *rondas campesinas* as a legitimate stakeholder in negotiations:

> "if you are their ally then they apply the politics of a good neighbor, social responsibility, etcetera but in the case of the ronda they not see them as a neighbor, but as an enemy and so with their enemies they apply very explicit security policies, in which the enemy has to be identified, has to be followed, has to be neutralized also right?"

YU used many warlike references and metaphors in his interview: while he says that the mine sees the *rondas* as the enemy, it is clear that he now sees MYSA in the same way. He perceives the company as using deliberate strategies, such as strengthening its relationship to the military and with the police, and having a "master plan" to divide the community.

YU asserts that "the company doesn't like democracy" and, like many other interviewees, identifies a clear alliance between MYSA and the government. Significantly he says that "…in the majority of cases it is difficult to know who the representative of the mine is and who is the representative of the government." When YU accuses the company of having two faces, he is contrasting the face of social responsibility and the face of repression, using police and military action.

Ways Forward
YU is at pains to point out that he, and the organization to which he belongs, is not anti-mining. The ideal situation would be for the company to not act in the ways described above: to not be repressive; to be honest about the potential impact of mining; and to be able to accommodate those sections of the community who do not want mining on their land.

2.3.5   NARRATIVE 2: FEMALE RESIDENT OF CAJAMARCA

YV believes that "from the beginning the mine has not had a good relationship with the people" mainly because they do not respect and value "the rural person": "it is not a good relationship because they do not value the human being, they do not see them as a human, they see them as merchandise."

Community Engagement as Gift Giving

"They are so desperate for the mega project Conga to go ahead that they are deceiving the people with books, with better kitchens, with shoes, with jumpers and so on. Some people accept the gifts and others don't."

MYSA as Deceitful

Like the narrator above, YV also perceives MYSA as being deceitful:

> "It is easy to deceive rural people, because they do not know how to read, because they are not aware about laws, because they do not know how the mine works, and so the mine takes advantage of that right? for example they don't inform them about the projects that they are going to do? For example, the Yanacocha mine said that they were going to be here for twenty years and well they have been here longer than twenty years, they lied to us saying that they are going to close the Yanacocha mine, and now they said that they are going to start another mine which they named Conga."

In the same vein, she believes that the villagers impacted by the mecury spill at Choropampa were "lied to, they hired doctors to tell them that they are going to get better, that that the mercury will leave and you know what the first thing that the doctors used to prescribe to people affected by the spill was "drink beer, that will take it out of you."

Lack of Support from Local Government Authorities

Like many interviewees YV sees a direct collusion between MYSA and government. In this instance, she refers to local government:

> "many authorities have sold their conscience for money like, for example Mrs X during the Fujimoro government was a council leader and she asked the people affected by the spill not to expose the company. She said she was in charge and she gave them assurances that she would get a fair compensation to all of them but nothing ever happened. That was another lie because she worked with, colluded with, the Yanacocha mining company. It has been confirmed, confirmation after confirmation that the government, buys authorities together with the mine, right? For example in the case of the governor … from here Cajamarca, at the beginning he came out to say that he was with the people and that it was true that the mine contaminated [the water] but more or less three days later he changed his stance, he changed his stance and he began to defend the mine and even now he is defending the mine."

2.3.6   NARRATIVE 3: FEMALE RESIDENT OF CELENDIN

YW is a middle aged woman, who identified herself as someone who was concerned about "the protection of our very unique water system."

Encounter with MYSA

YW frames the mine's relationship with the community as gift-giving: "they create relationships with gifts because in reality they have so much money, right? They take advantage of the naiveté of our people …." YW claims that the *campesinas*, who are poor, were susceptible to initial overtures of gifts, in the form of commodities or money.

She acknowledges that MYSA held many meetings with local people—"there were contingents of people going up into the highlands and they took people in their vans from the surrounding area, anyone who wanted to go." But she also believes that some of the meetings were skewed, that is, the company bussed in large numbers of their supporters and wanted to bar "trouble makers" from their meetings:

> "When they had their meetings here in Celendin my husband did not miss one and he documented what they said. I remember that before the meeting they would come and ask my husband not to stir up problems with so many questions. They said to him we can discuss the project with you in private but most people would not understand."

She is clearly sceptical about MYSA's motives in holding these public meetings: "they told them the good part, the nice part, let us say the investment part but the information about the destruction [of the environment], the poisoning [of the water] that is not discussed. They said absolutely *nothing* about that." She describes the reaction of the people at these meetings as "confused." "Lots of people were scared and concerned because they saw Conga—so beautiful, so precious—and they said 'how will it be?', 'who should we believe?' It was a new topic for everyone but most people were confused."

She claims that the purpose of the Thursday Dialogues, which were community consultations, held in Celendin, were to "brainwash people."

Changes in the Relationship Over Time

She believes that over time, people have come to hold a more negative opinion about MYSA:

> "Because to be honest those people who sold their land have realized that they are not better off, also we have been attacked by the police, the army, the whole government is against Celendin. They treat us as if we were animals."

She said that in 2013, a rural community in Cajamarca were in despair about the quality of their water supplies and it is these sort of communities who now realize they are not better off since mining began: "does that not make these men from Yanacocha repent?"

Response to MYSA

YW and her husband's response to mining is to lobby for the protection of the environment region (particularly the water resources) and to assist the *campesinas* who have been negatively impacted by mining. Water is a particularly important issue:

"Here in the Andes we care about water, it is more important than gold to our survival. Two-hundred and thirty communities will be affected, on top of this there are eighteen channels of irrigation, it is six-hundred and sixty-five of, them, there are also lakes. We want the company and the government not to touch the water basins. We also want them to control the use of cyanide, it is ninety-thousand tons of poison, and we will have to deal with it. They are going to kill us worse than dogs, worse than animals?"

2.3.7   NARRATIVE 4: MALE RESIDENT OF CELENDIN

YS is from Celendin but studied Economic Sciences at the University of Cajamarca. He has returned to his home town and is working for the municipal government. He only began to be opposed to the Conga project when the impacts of Yanacocha began to be felt. At the start of the mining project, he said "we were not aware of the impacts that mining projects generate."

Encounter with MYSA: Thursday Dialogues in Celendin

In fact, he observed that many local welcomed the largesse of MYSA in the early days of exploration and community consultation:

> "they arrived at schools with uniforms, with backpacks, educational materials, with gifts for the teachers; in the communities during their festivities, they gave fireworks, music bands, sport games, balls, do the people were used to sending a document to the mining company and the mining company simply gave something. So here in Celendin we used to say 'it is so good that the mining company is here, it is a blessing from God'."

When the Thursday Dialogues began in Celendin, the MYSA representatives (who had "something to do with Social Responsibility") would tell them "there is going to be work for so and so people, there is going to be a mining tax, and going to be a series of development projects, etcetera etcetera and with the gifts they used to give, we used to say 'wow, we are set'."

He noted, however, that responses to queries about environmental impacts were patronizing or overbearing: "are you geologists or environmental engineers?" they would ask. "But we had some knowledge. They behaved in the same patronising way with the communities when the people of the higher regions began to question them, they immediately stopped them, and they even ridiculed them in public."

YS describes the initial process of gift giving as a "fictional relationship" to get the trust of local people. He believes that MYSA, at least then, had deliberate strategy of dividing organizations, of financing the political campaigns of mayors who, when in power, facilitated the mining projects. Later, he was part of a contingent that traveled to Lima to lodge a document to the Newmont office. He said that "Newmont did not want to receive us. They closed the door; they brought in the police. Thankfully we had the support of a Congressman who communicated with them, then they allowed two of us in … we went in to give them our document and that was it." They also presented a document to the World Bank who was more welcoming but have not responded to their concerns.

Water, Land, and Knowledge: Contentious Issues

This type of response marked the beginning of a breakdown in communication between MYSA and the community, including both rural *campesinas* and urban residents. The water issue connected both groups of people, and is also very important to YS: "we realized that the mining company had located itself where the rivers are born, the same rivers that give water for agriculture, livestock, for human consumption of the communities, so that is our most important concern." YS went on to detail how much rock would be removed each day for the 17 years of life of the project—"facing this, we believed that we have to express our concerns."

YS also believes that MYSA used underhand means to acquire land, that is, they acquired signature of community leaders to authorize the sale of land which was communally owned. He claims that they then "told the people that they were practically illegally occupying their own property." YS cites the case of Narrative 6 (he knows the family) in which he portrays the company taking a threatening attitude to the family: "look you have to sell, you are going against the mine, the mine has influence with judges and the public prosecutors, we are a giant and you are ants." This incident "clearly shows that is them against us."

The way in which MYSA conveys information about mining, and the manner in which the EIA was delivered, are also contentious issues. YS says that the company representative use "technical terminology" that is difficult to understand that they make claims which are supposedly based on scientific fact, and therefore incontrovertible. "they told us, for example, that these waters from here, from the lakes, they are not fit for human consumption so they were telling us that the water we drank, that we had been drinking for generations and generations was not good and yet we have not died." The EIA for the Conga project was released in 2010. According to YS, this report was only available online from a natural resources and environment office in Cajamarca, not Celendin.

MYSA also called a public meeting to discuss the EIA but the meeting was held in San Nicolas de Chailhuagon which, while somewhat affected by the Conga project, is some distance from the region where the majority of people are affected, i.e., Sorochuco. YS says MYSA wanted to hold the meeting in San Nicolas because that community was pro-mining and they knew the distance would deter participants. Only a few people from Celendin went to this meeting, after hiring a minibus and traveling for five hours. People could write their names down if they wanted the opportunity to speak but could do so for only one minute. He said there were almost "a thousand DINOES patrolling the meeting."

Companies and Governments

Despite the government creating an Office for Dialogue and Sustainability to deal with conflicts like Yanacocha, YS does not feel that relations have improved. He cites the police presence in Celendin (in July 2012) as an additional catalyst to anti-mining sentiment. "They killed four of our men here in Celendin and one from Pampamarca and the shots came from helicopters and that is what it indicates, the post mortem of one of the men, that a bullet came from the air. So when they say "in Camjamarca there are four radicals that oppose the project" then what, to control four radicals, they come here with a big squadron of police and with helicopters? No there is something else happening here." The "something else" refers to the perceived alliance between the company and the government and YS says that the government "always supports the companies because the companies finance their political campaigns."

Ways Forward

The companies should respect the decisions of the communities and they should "accept there are some places where they can carry out mining ... but also accept that there are some places where mining cannot be carried out, they have to respect the communities." M.S. also feels that the company should not impose its idea of development on the communities: "before we used to live well, and it's like that they say 'no you are poor'. Lies with the discourse that we are poor, the mining companies come and impose their vision of development for us."

2.3.8   NARRATIVE 5: MALE CAMPESINO OF CELENDIN

YR did not speak about his personal experience of mining but about the relationship between the *rondas* and the police and special forces in Celendin, whom he sees as allies of the mining company.

The Celendin *rondas*, in his account, were against the extension of MYSA activities from Yanacocha to Minas Conga. He said that the government accused the *rondas* of "destroying the peace of the city" during the state of emergency at that time. The *rondas* main complaint was about the behavior the soldiers who were present in Celendin, at the behest of the government. According to YR, these soldiers allegedly sexually assaulted young girls; never paid for their meals in restaurants; carried out combat exercises in the streets; and that their encampment near the reservoir was unkempt, they urinated in the reservoir.

It was because of this, he said, "... the people were coming all the time to us the 'ronderos' and we had to receive their complaints. Everybody was fed up with the company and with the soldiers so as we are 'ronderos' it was our responsibility to take it up with the local authority and the police and we provided evidence of what the soldiers were doing." When it became clear that the solider were not going to be sanctioned, the *rondas* eventually "decided to act." They staged an intervention, that is, they caught soldiers who had been with underage girls and "told them off." YR is careful to point that they did not beat them or treat them inhumanely but delivered a lecture on their behavior.

Overall, YR feels the company "colludes with the public prosecutors, with the judicial powers, also the police, they are on that side." He goes on to say that "the authorities, the government and the company are dead against us, the *rondas*."

2.3.9   NARRATIVE 6: FEMALE CAMPESINA

YC is well known because of her, and her husband's, refusal to acknowledge that MYSA has legal rights to one of their plots of land which would fall under the Minas Conga mining lease. She claims that her family had no contact with MYSA prior to 2010, and no knowledge about plans to mine on their land as part of the Minas Conga project.

Encounter with MYSA: Machines and Violence

She says that "the first time that we were in contact with them was when they came with their machines to my property because they came to my land without authorization." She and her family went to MYSA's office in Sorochuco, only to be told that MYSA now owned that plot of land. This was confirmed by the MYSA office in Cajamarca who claimed that her father-in-law, Samuel Chaupe has signed a document which transferred the land rights to MYSA.

It is impossible to know whether in fact MC's father-in-law may have sold this plot of land or not. What is clear is that YC was not aware of these dealings and attempted to seek some kind of redress for what she saw as unlawful encroachment on her land by the mining company.

Seeking Redress

She therefore took the following steps: making a complaint to the police in Sorochuco; who then contacted MYSA without a response, and who then directed YC to "make a complaint before the public prosecutors in Celendin."

The family travelled to Celendin (see map) and explained their predicament to the office of the public prosecutors; a representative told them that it would cost money for him to come out to their property to "make an inspection." They agreed and borrowed the fee of 1,500 soles. It is not clear, to an outsider, what a physical inspection would prove when compared with, say, an investigation into MYSA records of land purchases compared with local records of land sales. In any event, the attorney never called YC or gave them a report on his findings. They later discovered that although their complaint had been archived in the Office of Public Prosecution "…all complaints expire after a certain time so all of our strength and money went in vain…."

Three months later, the MYSA diggers arrived at their plot again, this time accompanied by the police, another attorney from the Office of Public Prosecution, an MYSA lawyer, the engineer Silva (who seems to have been responsible for community engagement), and the MYSA security force. The state attorney advised them to make a deal with the company. When YC's husband said they wanted time to organize the move, rather than leave immediately as advised by the attorney, one of the MYSA engineers said "if you do not leave today, we will see that you rot in gaol." A violent encounter followed in which YC's daughter knelt in front of the digger, only to be forcibly removed by the DINOES who "dragged her by the hair, they kicked her, and they beat her with the butts of their rifles." The police prevented YC from helping her daughter, shots were fired, and she says they "took our food, clothes, tools, they killed our dog… before they left, they burned our ranch." When YC was at the property the following day, the police came again accompanied by Captain Soto (*it is not clear whether he is a policeman or works for MYSA*) who got angry with the police for the destruction of the Chaupe property.

Ways Forward

On reflection, YC believes that the "the mine does everything secretively" and that "the municipalities are just as guilty as the mine because they had not informed us that the mine will come to our community. They had not informed us about that at all." A friend of the family stated that the "the company should behave in a more civilized manner and should speak to the people. They should explain what the aims of their work are, and what is going to happen to the people, what are the consequences of the work they will carry out. They should listen and respect what the people want."

2.4   THE TINTAYA MINE SITE

The Tintaya copper mine is situated in the Espinar municipality in the southern Andean region of Cusco. Local people are predominantly of Quechua descent and, as with Cajamarca, farmers rely on clean water for their livelihoods. The mine was established in 1985 as a state enterprise and was eventually bought out in 1996 by BHP Billiton (BHPB). By 2002, local communities had presented a list of grievances to local, and then international NGOs, which included: complaints about the process of land acquisition; pollution of land and water; and sexual assault of women by local security forces. The Oxfam Mining Ombudsman, together with Peruvian NGOs, organized a Roundtable Dialogue framework which held the promise of positive outcomes but which were nonetheless seen as ineffectual by local people who participated in it. One thousand people stormed the mine site in 2003; although compensation agreements were signed in 2004, thousands more attacked BHPB facilities in 2005. BHPB sold its shares in Tintaya to Xstrata Copper in 2006. Two people were killed and 50 injured in large protests against Xstrata in 2012 followed by a state of emergency being declared a month later. Glencore International acquired the Xstrata PLC to create the newly named Glencore Xstrata PLC in May 2013.

The following timeline sets out the escalating conflict between BHP Billiton, then Glencore Xstrata, and the local communities. This historical sequence will provide a contextual backdrop for the interview material which appears in Table 2.2.

2.4.1   NARRATIVES ABOUT THE EXPERIENCE OF MINING AT TINTAYA

Narrative 1: Male leader of local Quechua community (TC)

Table 2.2: Timeline of Tintaya mine, and conflicts Peru (*Continues.*)

Date	Event
1985	Mining commences in Tintaya as a state-owned enterprise.
1996	BHP Billiton (BHPB) acquires the majority share in Tintaya and expands the mine by purchasing land from Quechua communities.
2000	CONACAMI requests Oxfam Community Aid Abroad to take up Tintaya case with BHPB head office in Australia. A coalition of five affected communities created an alliance with a group of domestic and international NGOs to build a case against BHP Billiton.
2001	Oxfam Mining Ombudsman visits affected communities.
Feb., 2002	First of many meetings of *Mesa de Dialogo* (Dialogue Table)
April-May, 2003	Frustration at the inability of Dialogue Table to generate tangible solution to grievances; 1,000 inhabitants storm the mine site.
June, 2003	Community members accuse company staff of not listening to them during the investigations of the commission or preventing women in particular from speaking to the consultants. Allegations that company officials have intimidated community members by stating that if they are involved in the Dialogue Table process then they are acting "against" the mine and will therefore not obtain work at the mine.
2004	BHPB and the five communities sign an agreement compensating families for lost land and livelihoods and establish an environmental monitoring team and community development fund (Acuerdo Marco Fund—referred to as the Marco Agreement in narratives).
2005	BHPB staff believe that relations have improved but thousands of people attack BHP facilities.
2005	Tintaya employees march in support of the mine in Cusco and Arequipa.
2006	BHPB sells share in Tintaya to Xstrata Copper. Xstrata agrees to honor the Agreement signed by BHPB. Tintaya pays 3% of its pre-tax profits into a community development fund as stipulated by a previous treaty with BHPIn theory, this fund is supposed to be jointly managed by the community and the company but in reality is managed by the company trust fund. Communities feel the agreement should be renegotiated.
May, 2012	Two people killed and 50 injured in protests against Xstrata's copper mine.
June, 2012	President Oscar Humala imposes a state of emergency in Espinar.

Table 2.2: (*Continued.*) Timeline of Tintaya mine, and conflicts Peru

April, 2013	Ministry of Environment released the summary of results of its Participatory Health and Environmental Monitoring (PHEM) which was commissioned by Peru's government in 2012 following violent protests of the previous year. The PHEM Report determined that there is pollution in the Espinar Province that appears to be the result of mining and there is pollution in the Espinar Province that appears to be from "natural" sources. Communities directly affected by Tintaya are exposed to lead, thalium, and arsenic.
	Glencore Xstrata responded that the contamination discovered above environmental standards was only in a few samples and that most of those samples were from outside of the "mine's area of influence"—asserting that the contamination measured was the result of natural, or "background," metals contamination, and not from current or Xstrata mining activities.
May, 2013	Glencore International acquired the Xstrata PLC to create the newly named Glencore Xstrata PLC.
June, 2013	Publication of a report on environmental impact concerns prepared by Centre for Science in Public Participation, at the request of Oxfam America. The report is based on review of publicly available data and reports and visits to the communities and the mines (tours provided by Xstrata personnel). Also in this year mine closure begins—to be complete by 2039.
2014	Xstrata Tintaya fined $84,000 for pollution (elevated levels of copper in soil).

Narrative 2: Female resident of local Quechua community (TH)

Narrative 3: Female resident of Espinar (TE)

Narrative 4: Male farmer (TF)

Narrative 5: Male living close to mine site (TA)

2.4.2   NARRATIVE 1: MALE LEADER OF A LOCAL QUECHUA COMMUNITY

Encounter with Tintaya

This community has lived with the mine for 30 years. According to TC, the community's first encounter with the project came when the company (then owned by the government) said "we are going to build a road" and "so the community approves, the community accepts the road will be built, that it will pass through the community. But after the road, what happens next? Then the mining pipelines come. They betray our trust, why do they not speak the truth and say what will happen to the communities?"

He also said that previous community presidents sold land to the company without informing the local people: "we found a document signed in 2009–2010 selling land to Xstrata Tintaya but he (the ex-President) did not inform us about these documents. He was doing them under the table with the authorities and the company and the community never even had the knowledge about these documents that were being signed and registered by the notary."

Deteriorating Relations Over Time

He used to think of the mining as being "respectful" toward local people but now "they work with the intelligence service, with the police" and treat people who question mining practice as criminals. When asked about specific personnel who engage with the community, TC said that "we do not know of anyone, an engineer or a representative with whom we can speak directly. There are managers, there are supervisors, but they do not have their doors open for dialogue, to speak. There is no one we can trust."

His overall feeling about the company is that of betrayal: "they have never officially consulted me or notified me of what they were going to do. We feel betrayed by them. They do not inform us clearly that they are going to do, they don't discuss their plans, and how they are going to do it. Indigenous people do not know how the mine works, so they only shut us up, they only humiliate us and that is how the people are living." In continuation of those feelings, he said that he—and others—feel abandoned by the Peruvian government.

2.4.3   NARRATIVE 2: FEMALE RESIDENT OF LOCAL COMMUNITY

Treatment by the Police After the 2012 Demonstrations

TH is a Quechua woman, whose family speaks Quechua (rather than Spanish) at home. When recalling the 2012 protests, she says:

> "We have been beaten, by the police. We have been treated like they would a piece of cloth, they just threw us in a corner, then they kicked us, they hit us with their batons. They beat my husband too, they beat him, and the police dragged him. It seemed like we were rags, which is how they threw us about. My husband is the president of our community but the company doesn't like him and they say they will take him out although he is the elected president the company has named another person. They want a president who can work in their favor, nothing else."

Perceptions of the Company

The perceptions of the company were also impacted by their actions toward the local community as indicated by TH:

"The company doesn't want to talk to indigenous people, they should speak, they should come and dialogue, but they do not come, they don't talk to us women. We want to learn. I belong to a mother's club and we are worried about the children, they are often ill with skin irritations but they never been to the club to talk to us. My husband says we cannot talk to the company because of the way they have mistreated us. They have mistreated us too much. That's all I have to say."

2.4.4   NARRATIVE 3: FEMALE RESIDENT OF ESPINAR

TE has a family, and is a member of a women's NGO. She was a teenager when the Marco Agreement was formulated with BHPB. As an adult she was approached by the company (now Xstrata) who said to her "because you are female and you understand the Marco Agreement, what is it that you want? So I asked for projects for women." She advocated for programs which would help women's livelihood such as craft production or animal husbandry. TE did not feel however that she should have to go on radio, as requested by the Xstrata Tintaya Foundation, to publicly thank Xstrata for their funding: "I could not say thank you, because no, that is not their money, it is ours."

Community Engagement: Image not Substance

When asked about community engagement, TE recounts how she is often told by company relations people that "Of course the company has changed," they say, "we have changed, we are working side by side with you, we have changed." The change, however, is exemplified in the distribution of gifts and cash, ie in compensation. But for TE, this type of compensation is not always suitable for Quechua people. "For example, they are giving people tractors with that money from the Marco Agreement. The townspeople, they do not work the same way they used to, how it should normally be done. It is all mechanical and mother earth (*Pachamama*) is not well, it is not matured enough and now it does not give fruit as usual."

She raises these issues on the radio and discusses, for example, the lack of pasture and water for people to keep their animals: "I know that I am not welcomed, I am being accused of spreading false rumors about the company during my radio programme. Someone from the community with the support of the company filed a complaint against me and I have my first hearing to be informed of charges on the 5th of March."

TE feels that the company manipulates community relations and is more interested in producing an image of good relations rather than genuine dialogue. She provides an example of the company wanting to hold a street party to celebrate the 10th anniversary of the Marco Agreement: "what the company wants is to decorate the streets, they prepare a special meal and ask the women to dance and during all that time the company is recording everything and then they say to other communities, look at the people of Espinar are happy, look at the women dancing! But we are not going to let it be, we have come to an agreement that we will not participate in those exhibitions, we are not going to participate in any programme."

Her resistance to this representation comes from a belief that the company is not following the Marco Agreement: "it has not been followed, for example, the people from Espinar should have had work, but no, it was not like that, the workers of the company were all from other places, from other towns, so when they receive their wages, they do not consume things in Espinar, but in other towns. But in Espinar, as the mining company is there, the living costs are high, as if we are all working in the company, but it is not like that."

The Marco Agreement, she feels, has generated a cultural shift amongst many local people who are now willing to settle for hand-outs:

"They have taught people to ask. For example they do not want to work anymore, but, if someone from one institution says they have gifts then they all turn up , but if they are going to give information about the environment, or about the monitoring which tells how the water and land is, then no one participates. If the mining company is going to talk about the Marco Agreement, and it is announced that Xstrata Antapacay is going to be present then yes, the people attend, but when it is an NGO, no, they do not go to see."

People who do actually question the company about environmental issues, such as contamination and waste, are dubbed "anti-mining." If local mayors want more information about certain issues, they are accused of "holding back the town." For TE, mining and cash hand-outs have divided the community:

"Nowadays money is the only thing that people see, money, where is more money?"

Ways Forward

TE would like the company "to work transparently as the law states but they do not respect that." But if the government does "make the company respect" communities, she wonders how this will happen.

2.4.5   NARRATIVE 4: MALE RESIDENT CONCERNED ABOUT FARMING

First Encounter and Gradual Exposure to the Company

TF first visited the company offices in 1995 (after BHPB took over) with other community members. They requested about 5,000 gallons of fuel to "open a pathway to the settlement" and the company said "yes, no problem but they also said "you have to give thanks" so I thanked the company (to be honest we *were* grateful) in an interview which was recorded and later on it was played by a radio station." Then in 2002 after he gained a position representing a group of farmers in Espinar, he came in closer contact with the company and experienced first-hand "how they worked, what benefits there are for Espinar, and their ways of working."

Company Ways of Working: Reward and Punishment

TF feels, like TE in the narrative above, that the mining company has "divided the Espinar population with their economic power … by now they have about 20 radio programmes in the Espinar province." They will offer to put on radio programmes, for example, or fund community newspapers.

> "Because of the money, sometimes the people accept, sometimes even if it goes against the population, against the authorities, sometimes we are willing to sell our conscience, that is what is happening."

TF says there is culture of surveillance of people who speak out, not even to stop the mine but to question its activities. He cites the example of a provincial mayor now in his second term of government but he is accused of not being co-operative, or of being difficult. The Marco Agreement is under re-negotiation but TF is not optimistic because he feels most of the local mayors are "biased in favor of the mine."

He also feels that relations have deteriorated:

> "In my opinion the mining company has no incentives to improve. The company is used to get what they want. I think it will be difficult for them to improve in anything. The company rules and… people are used to it. The Marco Agreement is currently in re-consideration, but the mayor of the Espinar province, the mayors of the districts are biased, they are in favor of the mines because the district mayors get offerings by the representatives of the mine: 'what do you want for your district? … because they have money, people request it, and the company can then say 'the population is in favor of the mining company,' 'it gives support,' but we have to be careful with that and that is what is happening now."

Water Issues

TF emphasizes that agriculture and cattle-rearing are the mainstay of the local economy. He is concerned about the contamination of waterways for both humans, plants, and animals: "the mine is currently at the water heads of the rives, where the water flows down, in that river you cannot bathe in any more, when you bathe you start to feel itchy, you get a rash on your body, many things start to appear, there is not, we have managed to prove that the is contamination, but they say it is not."

On the Tintaya Dialogue

> "I had the impression about the representative of the company, I do not know but I felt like, they laugh saying 'of course, yes, we will do it' but they do not value the discussion. They are always talking about the price of minerals 'the price is down we cannot increase our contribution.' Under pressure they say 'we approve it,' they approved one thing and then another, but when the time came to sign the act, they

did not want to sign it, so it was just a waste of time the dialogue. That is the way that the mining company Tintaya works, what can we do if they do not want to sign the act?"

2.4.6   NARRATIVE 5: MALE RESIDENT LIVING CLOSE TO THE MINE

Land Negotiations

TA negotiated an initial agreement with the company (then Xstrata) about the sale of some of his land. He said "it was a bad price" but "in the second negotiation the company came to us, they came with fizzy drinks and their bread to practically gain our friendship and our trust … they talked about living well together, being good neighbors, but the company practically changes engineers, co-ordinators all the time so we do not even know the people there, right?" He has another parcel of land the sale of which was not complete:

> " … we returned by day to the plot of land that we were negotiating, and in the afternoon the mining company came with their killers, the police and security guards that was at night time, they practically surrounded the whole family, they practically threw us out around eight in the night, nine in the night, during the night, and together with the cattle we had to go, they evicted us, right? So they displaced us … some animals were left behind because it was night time … they came and trampled all over us with the pickup trucks, with the police, disguised with balaclavas. So the company behaves in that way, we had the just reason, that if they did not meet what they had promised then we had the right to return, right? … The lawyers, the public prosecutor they all were there telling us 'you sold the land'."

Living Conditions

The land which TA owns is very close to the mine site. While this makes it desirable for the company to purchase, it also makes it an undesirable place to live. His land is only 7 meters from the mine site: "the noise is very intense, it is too much, too much," then there is the dust in the dry season which is affecting both humans and animals. He says the water supply is contaminated but when they complained about it, the company promised to "monitor the situation" but they have not received any reports, and are still drinking the same water.

2.4.7   NARRATIVE 6: FEMALE RESIDENT LIVING CLOSE TO THE MINE SITE

TJ is a Quechua speaker who lives close by a mine site. She and her family also sold some land to the company at a very low rate, not because she agreed to that rate but because the company got her mother to sign the deeds in her absence. Her mother does not speak Spanish. They advised her mother to relocate to the district of Coporaque but EJ says "this area is dry, there is no water, there no access to a road, the road does not go there, how will she get there?"

Water Issues

She is very concerned about the water supplies to her property. The company had built one canal but the water, she says, is not fit for human consumption: "my children have rashes, they have headaches, stomach aches. I have started buying water from Espinar town, in bottles for our weekly consumption." She does not want to leave the area however: "I am not going to go anywhere. I am staying and to shut me up they will have to kill me first." Her mother's experience has made her suspicious of company dealings. Many people in the surrounding area do not support TJ's claims so she has contacted the Espinar Municipality to get the water tested.

She feels attachment to the land even though it is difficult to live there, so close to the crushers and the continual dust. She is trying to continue living on the land as her father did before her: "he sowed clover seeds, but now I am trying to following his path, but every year I am sowing but every year I fail, there is no water, every year I fail, sow and fail, this year I have three hectares half sowed, plus one hectare, and two hectares of pasture, dormant and recently sowed, I do not know how it will end up this year because we do not have water and the mine, just walks all over us."

2.5   COMMON ISSUES

Analysis of the Yanacocha and Tintaya narratives revealed that both groups identified common issues which had shaped mine-community relations. These are listed below and include some quotes which have already been presented in the narratives, additional quotes from the same interviews, and quotes from academics and NGO personnel in Lima. It should also be noted that many issues overlap.

2.5.1   FLAWED COMMUNICATION PROCESS

Withholding Information

There is very little evidence, from these narratives and from other sources, that exploration companies or mining companies explain their goals to communities before beginning exploration or mining activities.

> It is quite normal in Peru for the state to decide to do a new extractive project or an infrastructure project without informing the communities let alone consulting them. They simply do not participate and people normally find out that their territories are in the process of being granted for concession or that they form part of a license when the company is already entering their land to carry out their extractive activities. As I said, this is a big problem because many of the communities in our country do not know that their territories are no longer theirs and that is an issue. The government totally bypass them.... [In the case of Yanacocha] these subsistence farmers have not had any knowledge that their lands were part of the Yanacocha expansion. This is important because the company is arguing that as it is not a new project they do not

need to obtain the consent of the communities. This is why the company is treating the Conga project as an expansion of the Yanacocha mine. As a result, the relationship between the towns and the company is very concerning. Instead of informing and seeking their consent the company is promoting handouts to the towns so that they can accept the Conga project (Interview with environmental lawyer).

The Cajamarca NGOs were also concerned about the mechanisms through which communities could find out about a mining lease within their territories, or how they were informed about when to submission regarding EIAs or changes to mining procedure. In Yanacocha Narrative 3, M.S. relays how the communities in Celendin and Cajamarca were informed about deadlines for comment on planned changes to water usage through an announcement placed in the newspaper "El Peruano" but many communities did not receive this newspaper or only received a few copies.

"Yanacocha published a note in the most widely circulated newspaper in the region. Yanacocha was asking for permission to use the water that we consume and they gave us until the third of January as the deadline to present any opposition…in the communities people do not read newspapers, so by what means were we supposed to find out? Is that fair?"

"The company comes with betrayal, with stories, we are going to build a road and with that tale so the community approves, the community accepts that the road will be built … but what happens next? Then the mining pipelines come. They betray our trust, why do they not speak with the truth and say to the communities what it will happen" (Tintaya Narrative 1).

The lack of information about acquisition of land, about who is purchasing what land and from whom, is a divisive issue:

"the mine …created allies with the Peruvian state, in order for them to be able to negotiate individually with indigenous communities. Now, how can they do that when the communities have collective rights? …in our country rural agricultural communities exist, and they share the land, collectively, to be clear , they own property in a collective manner, now for the mines to be able to enter and do what they did they needed the support of the government and so the state contributed to this with a new policy to give out land titles, a policy promoted by the World Bank …The way I see it, the communities did not participate in the decision making of the concession nor were they informed of the conditions or the process the company would follow for buying land. This lack of information is a big problem and divided the farmers in to two groups one that was paid large amounts of money for their lands and others who got less and felt cheated" (Interview, environmental lawyer).

Difficulties in Accessing Information

Much of the information about the mining process is disseminated in Spanish presented in long documents filled with technical terminology. According to a researcher for Cooperaccion, a Peruvian NGO that monitors mining:

> "In the case of Peru, for example when we speak about high Andean communities, we are speaking about Quechua-speaking communities, we are speaking about communities that the ILO identifies with specific needs for example their culture, their attachment to the territory, own language so, I can't give these communities a document of twenty thousand pages, which is what the Environmental Impact Study is, and say revise it within twenty days, and give me your opinions, when that information does not arrive in good time, is not adapted to their culture, to the reality of the communities, so that at the end the information is not useful. Personally I would say that the delivery of information is … also incomplete and so, that prevents any public participation process, or participation process of the communities in the areas of influence, it is a participation process that is very limited, and that has all of those limitations …it is a participation that is not informed, and that is not conducted in good time frames, so then it does not influence the making of decisions of a mining project" (Interview academic and NGO member).

The narratives also reveal a deep frustration with being uninformed, to the extent that people feel this is a deliberate process of marginalization. In Yanacocha, for example one informant complained that "now they come with this tale that they are going to bring electricity pylons through the community. We were never informed about these mining pipelines, no the people in the community were not aware of this." In Yanacocha Narrative 5, YV says it is "easy" to deceive rural people.

Not Appreciating Local Ways of Understanding and Communicating

There is frustration with company personnel being seemingly unable or unwilling to communicate in ways that are direct, transparent and easy to understand. One member of an NGO organization commented:

> "[it's an issue of] methodology the engineer simply talking at them and well the people do not follow the rhythm, of those types of meetings and then, the people are sat down, they sign the act but they have not understood, and then when there is the stage of the questions, rural communities always go for practical questions, they do not work in situations where they are sat down, they are listening to the details, they go straight to very concrete questions, "is the project going to contaminate or not contaminate? To questions like that, they expect clear answers 'yes or no' so the answers they get do not satisfy them" (Interview, NGO Lima).

There is also a common view among those interviewed that verbal agreements are binding:

"so we complained, 'what is this, we have some of the agreements we had come to, that we are going to work and that you are going to give each one of us work' and practically they did not do that, 'well where is the document that the engineer signed?' Yes, that is what they told us, but we only talked, we had trust in the company, that this was going to be, and we thought that we had a verbal agreement, we thought that they were going to do it, however the company at the end of the day they acted like they did not know anything" (Tintaya Narrative 5).

2.5.2   SCEPTICISM ABOUT COMMUNITY ENGAGEMENT

Giving Gifts as a False and Empty Gesture

There are many accounts about the questionable largesse of mining companies in the early life of a mining project.

"they are so desperate for the mega project Conga to go ahead that they are deceiving the people with books, with better kitchens, with shoes, with jumpers and so on. Some people accept the gifts and others don't (Yanacocha Narrative 5)."

"[The company] they are giving people tractors with that money from the Marco Agreement … hey have also given, cows, alpacas to the farmers…for example if there is no pasture in that community, if there is no water, how are they going to keep their animals?" (Tintaya Narrative 3).

"they arrived at schools with uniforms, with backpacks, educational materials to the teachers, with gifts in the communities during festivities, they gave fireworks, music bands, sport games, balls so the people were used to sending a document to the mining company and the mining company simply gave something" (Yanacocha Narrative 7).

This type of gift giving is not seen as generosity but as empty gestures to buy community goodwill.

Giving Gifts as Corruption

There is also a perception that gifts are also used as bribes:

"we found that they said that 580 people participated in one communal assembly and 515 at another, what they have is signatures they obtained from distributing food, bread, sweets, that is how they got the signatures that made them the legal proprietors of the land. The previous community presidents, have never informed their people of these documents which means that they too acted against their own people. We also found a document that apparently was signed in 2009/2010 it has 580. This time the community was selling land to Xtrata Tintaya. We have seen this document. The thing is that the ex-president did not inform us about these documents that he was legally doing. He was doing them under the table with the authorities and the company and the community never had the knowledge about these documents that were being signed and registered by the notary" (Tintaya Narrative 1).

Empty Rhetoric

Many people affected by both Yanacocha and Tintaya are deeply sceptical about the process of community engagement. In Yanacocha, for example PS (Narrative 5) conveys his perception of community meetings:

> "the mine …fills the room up with people that are in support, even with their own workers, the people …affected, they do not allow them in …what is important is that during these events there is no possibility to dialogue, instead , instead people can write down in a piece paper [about a] problem …but they do not receive an answer and the study …it is approved so it is only a formality …. There have been many aggressions against people who have tried to get in …the police beat them up and stops them from entering. Now it is even worse …they have divided their projects into smaller parts in order to avoid these hearings …now they can jump over them and the population simply does not have the possibility to dialogue with the company until they see the machinery in their land, like the case of YC."

TE, from Tintaya (Narrative 3), is also scathing about company public relations which, she says, always wants to depict laughing, happy Quechua people in their promotional literature. MP, who works for a human rights umbrella organization in Lima, says that " …companies treat communities like children, they give them small gifts that in reality do not compensate the harm that the extractive industry is causing them, but they do maintain them linked in a clientèle type of relationship" (Interview NGO Lima).

Deterioration in Community Relations

Considering the escalation in social conflict around both mine sites, it might seem to be stating an obvious fact that people feel relations have deteriorated. It is the reason for that perception that is important, however:

> "in my opinion the mining company has no incentives to improve. The company is used to get what they want. I think it will be difficult for them to improve in anything. The company rules and …people are used to it. The Marco Agreement is currently in re-consideration, but the mayor of the Espinar province, the mayors of the districts are biased, they are in favor of the mines because the district mayors get offerings by the representatives of the mine: 'what do you want for your district?' …because they have money, people request it, and the company can then say 'the population is in favor of the mining company,' 'it gives support,' but we have to be careful with that and that is what is happening now" (Tintaya Narrative 4).

2.5.3   COLLUSION BETWEEN COMPANY AND GOVERNMENT

Use of Police and Special Forces to Protect Company Interests: Pivotal Moments in the Deterioration of Mine-Community Relations

There is agreement among those who are disaffected by mining projects that the government supports the mining companies and protects the interests of the company. This is most deeply felt by the use of police and the Special Forces (DINOES) to enforce company policies.

This is evident in all the narratives but particularly the following:

- story of forced eviction in Yanacocha Narrative 6

- portrayal of the police presence in Cajamarca in Yanacocha Narrative 2

- description of police treatment in Tintaya Narrative 2

- account of forced eviction Tintaya Narrative 5

- account of police treatement in Tintaya Narrative 6

The use of force in response to social protest has proved to be a pivotal moment in the deterioration of mine-community relations. The death of protestors in Cajamarca, for example, was perceived in the following way:

"It was a pivotal moment as it was the first time that someone died in a protest …also it is the first time that there is the use of arms, additionally it was identified that the company had firearms and that opened a new issue …Also it was shown that they had a plan …for the military to control the area …the operation was called Operation Devil …so the company instead of taking measures to improve the quality of the environment, or improve their operational capacity to control their operations, what they are doing …is looking for clearer repression plans in alliance with the police, …with the military" (Yanacocha Narrative 3).

The use of police, in addition to mine security forces, is troubling to local NGOs:

"the companies still have their agreements with the police …Yes, there are various companies in the country that are operating with their own security group but there are policemen paid directly by the companies, they eat, sleep and mobilise using the resources of the company, they also receive direct instructions from the company. They are rented police, ordinary policemen with their regulatory weapons, with their uniforms, but that are doing private services for the company. This has been official since the time of Fujimori, but since 2011 policemen can lend their services to a company at any time. When it started it was only when they were in their holidays, now it is any time. So for example they may need the police to take care of the streets of Lima but it doesn't happen that way because Yanacocha needs them and so this generates a vulnerable situation for human rights, because the police is sent by the companies to very isolated areas, and you are putting public forces at the service of private interests, it also is a very discriminatory issue, because at the end of the day the one who receives protection from the state is the one who can pay for it.

Here in Lima, and well in other parts of Peru too, we have a very serious public security problem, and now the case is that the policemen who had been trained with our taxes, paid with our taxes, are going to work for the mine, to receive an additional wage. Because it is an additional wage the policemen go there very happy because as the state pays them very little, they want to go where the company needs them and the other problems is that, it is not just that the police go to the communities to work and comply with the constitutional regulations in the middle of no-where, but the issue is that they receive direct instructions from the company, and it is because of that the people have died, for example in Cajamarca …for example in the area where YC lives, probably the only contact that she has with the state is with those policemen paid by the mine, that is the face of the state for YC. They are the police that come and repress her, because the mining company has sent them, and that happens in many parts of the country, in the Amazon, the highlands of Piura, that is the face of the state for the people" (Interview NGO Lima).

The Government as an Ally of the Company

There is a shared perception that the government does not act as a regulator of mining, but as an ally of the company, working to serve the company's interests.

"The company has the government as an ally. During events they both appear together …It is difficult to know who the representative of the mine is and who is the representative of the government …sometimes the government representative talks more in favor of the company" (Yanacocha Narrative 3).

"I think that our government looks like they work mostly with the private companies …the government of Peru does not support the people of its own country. When a community complains, the central government supports the companies more with their police. It is the humble people in this town that live in humiliation" (Tintaya Narrative 1).

"There is no pint in talking to the authorities …they are also biased. I don't blame the company I blame the government. Since Fujimori the Peruvian laws have only been for the benefit of the transnational mining companies, and I believe today there is no respect for the people, there is no respect of human rights, that is what is happening in the Espinar Province" (Tintaya Narrative 4).

"The public prosecutor does not work on Saturday and Sunday in Espinar … only from Monday to Friday they work. But that Saturday they were there evicting us the public prosecutor is present there …but when a person from the community wants them, they always say they have a problem, are they going to come out on Saturday and Sunday for *us*? No they don't work weekends. But for the mine the authorities here in Espinar lend their services, and this is concerning for us, definitely, so that is what is happening and since that date, we do not have a relationship with the company" (Tintaya Narrative 5).

The Government Designates Those who Complain or Protest about Mining as Criminals
The experience of forced evictions at the hands of mine security forces and police, and the experience of police brutality during the states of emergency at Celendin and Espinar, create fear and is associated with historic memories of the violence generated by the armed conflict between the Peruvian government and the Shining Path insurgency. MP, a resident of Lima who works with an umbrella human rights organization has this to say:

"those who oppose the mine are de-legitimised, they are also branded terrorists because here in Peru, with a history of armed conflict, that is something that weighs a lot, that terrorise people because, until very recent the terrorists here, or those presumed to be terrorists, they would take them out of their homes in the middle of the night, shoot them and make them disappear so it is a big stigma to be accused of being a terrorist" (Interview NGO Lima).

"if someone speaks in favor of their community …they are marked, that person is threatened, for example me, we have informed about these documents and already the company has blackmailed us, they have criminalised us, they even want to take our position …as president [of the community] because the company does not like that I talk about these documents. I should keep quiet: whoever informs the public, the company does not like them, they only like people that humiliate themselves, that they can be paid to shut their mouths. And we are living like that" (Tintaya Narrative 1).

MP explains that the police are the "face of the state" to the rural people in the Andes. The state is not seen as an ally, or a place where people can safely take their complaints or seek a form of redress. Some of the Tintaya narratives mention the example of a local mayor who is unusual in voicing concerns about the impact of mining on the environment and the community:

"The company says to the mayor, 'you have to agree with what I do,' so if the mayor agrees then they work well, but if the mayor says I want more information or he says the company is not doing well, then no the company does not support the municipality. At the moment they say Oscar Mollohuanca the mayor is holding back the town" (Tintaya Narrative 3).

Local individuals such as this mayor (and Marco Arana in Cajamarca[3]) advocate for the people, but are marginalized and seen as "troublemakers." The director of a human rights organization in Cuzco took a similar view:

> "But I think that the communities hope for the state to have a role, more to regulate and guarantee rights, but ...ordinary people have no trust in the government, in a way their association with the companies has weakened the state itself, and sometimes it is the cause of the conflict because people have nowhere to go so if a problem can't be resolved directly between a company and the communities and many examples come to mind and the state is an observer but a pro-mining observer, what can people do? Yes there is a legal framework, but what one can see is that in practice the legal framework is brushed aside and what counts is this big relationship between company and state" (Interview NGO Cuzco).

2.5.4   PERCEPTION OF MINING COMPANIES

Mining Companies Minimize Concerns About Their Activities

When local people express concerns about the impact of mineral extraction and processing on the environment, they feel their concerns are often trivialized or glossed over.

> "When you questioned the impacts that the project was going to have, they immediately came out in an overbearing manner to minimize and even put down the arguments that we were expressing, saying that why do you argue if we are not specialists in mining. Are you a geologist, or environmental engineer ...they behaved in the same patronising way with the communities when the people of the higher regions began to question them, they immediately stopped them, they even ridiculed them in public" (Interview 3: Milton Sanchez). "But the information about the destruction, the contamination, the poisoning that is not discussed, they said absolutely nothing about that" (Yanacocha Narrative).

> "we also spoke with a lady from the environment, I do not know what her name was, but we also spoke with her and the lady from the environment she pledged to give us a report about the impacts. The report would confirm if the water was good for human consumption. She promised to the send the water to the laboratory, and give us the results ...and they were going to give us a report in a week, until now I think that was about four, five months have passed and now practically we are still consuming the water, but the water comes oily and splatters out and that water we are sometimes drinking, they say that it is ok for human consumption, but we are killing ourselves in life .... We get ill a lot, it is mainly the children and they run a higher risk against their health, because they have longer to live but the children

---

[3]Marco Arana is a former Catholic Diocesan Priest who worked with an NGO group that monitored the impacts of mining in Cajamarca (Arana Zegarra 2004) [1].

some day they are going to suffer from who knows what, and that worries us .... We don't know what to do" (Tintaya Narrative 5).

"…there is a global tendency at the moment to promote the voluntary principles, they also talk about self-regulation, …that it is an attempt by the companies to stop the state from intervening, 'it is not necessary for you to regulate us, alone we are going to behave well' what I think is that the state should meet its obligations, which is to protect its citizens but unfortunately in countries like Peru, the state is too weak and does not meet its role" (Interview NGO Lima).

2.5.5   NEGATIVE PERCEPTION OF MINING EMPLOYEES

How engineers are perceived.   The narratives often refer to engineers although it is not clear what positions they hold, and at what level. Engineers were described as people who:

Hold Power

"the engineer Guillermo Silva was the only person who could solve my problems" (Yanacocha Narrative 1, describing how the Chaupe family sought clarification about their land).

"I haven't had any contact with the company apart from the time when the police came with the engineers and they attacked us. They are terrible" (Tintaya Narrative 2).

Are Indifferent to the Needs of Ordinary People?

"we went to the mine's office in Cajamarca and there the lady in charge phoned the engineer. He didn't show up …the engineer refused to give us an appointment" (Yanacocha Narrative 6, seeking clarification about their land).

"So I had been working there for a year, and then they were reducing personnel, so then I went to the engineer and I said "you know what Mr? I am a little unwell and so could you look for another area for me to work in" just as he had promised, so then he said "you know what Mr, in fifteen days I will look for a place in another area for you" so then we agreed on that, then fifteen days later, and nothing. So since then, I kept telling them and they kept saying in two weeks time and a month passed two months, passed, three months passed and practically, then, *he did not want to even recognize me*, until now practically I am wasting all my time waiting for them to do what they said they would do, right? waiting, but the years go by wasted, we are wasting time, we have a family, everything" …that engineer …does not want to know me, I mean they do not want to assist the people that have negotiated their lands, and it is those engineers that practically have never showed their faces *when you speak to them nicely, they do not respond to you, you call them to say hi, and they do not listen, they pretend to be deaf, that is how the engineers are*" (Tintaya Narrative 3 describing how his attempts to get a different job on the mine).

"the engineers they say 'go and complain somewhere else' like that they tell me 'if you do not want to move from here, then it's going to be like this', that is what they say" (Tintaya Narrative 6).

Community engagement (CE) personnel on the other hand are perceived as the "nice" face of the company. The Yanacocha narratives indicated a greater contact with CE personnel but although some admitted that these people were well meaning, they had no power and often did not stay long in the position (see for example Yanaocha Narrative 3). Engineers were also described as transient: "the company practically changes engineers …all the time, so we do not even know the people there" (Tintaya Narrative 5).

2.6   DISTRUST OF CSR

As discussed in the Introduction to this chapter, we also interviewed a small group of NGO personnel. There is general scepticism about the application of CSR (corporate social responsibility) in the absence of state regulation, particularly at Yanacocha and Tintaya as illustrated in the following quote:

"…That (CSR) was a strong discourse shared by many companies in Peru. However, what we have actually found, is that in reality the discourse has various problems, first …is the big difference between the discourse and the concrete actions. There …is a great abyss and an enormous gap. One thing is what the code of conduct says, the code of social corporate responsibility, and another things are the actions. A second problem with the discourse of Social Corporate Responsibility …is that this discourse attempts to replace the local public politics and to replaces the state as a means of control, for supervision or regulation. The discourse is the discourse of the auto-regulation 'we the modern companies are conscience of our activity, as we are referring to mining activity, it generates some negative externalities, some negative impacts in the populations, but do not worry, because with our social and environmental standards we are going to control those negative impacts'. So although the companies are talking about social standards, of environmental standards, and that they have a code of conduct and social responsibility, we do not think that it is sufficient. Because the state has to play a role, the state has to control, regulate and supervise, the state, let us say it represents the common interests, the common good, and in that sense the state has an important role …it is the absence of the state, the absence of the government what without a doubt provokes a completely and absolute asymmetric situation, and that is the scenario for which the conflicts …we feel that the absence of information is made worse by the fact that the state is absent" (Interview NGO Lima).

2.6.1   VIOLATION OF HUMAN RIGHTS

Overall, these experiences—of people who wish to question mining practices, or who do not wish to sell their land, or who oppose the extension of a mining project—constitute a violation of their rights as people who deserve to be treated with dignity and respect. It makes some of them afraid to pursue further action or to question company policy.

"I have complained [to the company] about the way they failed the community …the company has not liked our observations so now they slander, blackmail, bring the police, …they made us fight with the police. People are afraid, they no longer want to reclaim their rights, and they do not want to talk about their rights. One time when us leader complained about these [false] documents they began to threaten us with the police, to follow us, day and night we are being watched, we are running risks, even our lives are in danger by the large companies that come to our territories" (Tintaya Narrative 1).

On the other hand, it has made others desperate:

"I am not going to go anywhere. I am staying and to shut me up they will have to kill me first. You see I cannot allow the company to walk all over me, it is ok for them, they have money to get lawyers, they have five lawyers, policemen come, but me, I am practically alone defending what is mine" (Tintaya Narrative 6).

2.7   COMMUNITY SOLUTIONS AND GUIDELINES

In the interviews, we asked people if they could identify ways forward to improve community relations. Not surprisingly, the key change areas are the reverse of the negative qualities that currently permeate mine-community relationships: instead of paternalism, fear, scepticism, contempt, and violence, the people want a relationship that is founded on transparency, trust, respect, and peaceful negotiation.

Guiding principles for engagement, as defined by local communities, are summarized below:

- Transparency

- Respect

    - Respect for local values and knowledge

    - Respect for alternative models of development

    - Respect for human rights

- Dialogue

    - Effective dialogue should be inclusive

  – Effective dialogue requires a review of time frames

- Safety

  – To ensure the safety of communities, government needs to be an effective regulator of mining practice

- Capacity

  – The communities' capacity to participate in dialogue with mining companies and governments requires the fulfilment of all guiding principles listed above.

2.7.1   TRANSPARENCY

The need for transparency refers to the need for a clear understanding of the kind of work that is going to be carried out, how long it will take and what impact it will have on people and environment:

> "the company should inform us about the work that they are going to carry out within the community …and how they will live with the community …they should support the community but apart from that the mine should work on the environmental impacts …because the first need nowadays there is no water but the mine has never said to our community you need water so we are going to help you with that. They have never said that …they only worry about their own interest, their projects, their production and so on. They should support the communities" (Tintaya Narrative 1).

> "now they come with this tale that they are going to bring electricity pylons through the community. We were never informed about these mining pipelines, no the people in the community were not aware of this" (Tintaya Narrative 1).

2.7.2   RESPECT

The idea of respect covers many domains. Every person who was interviewed wanted to be treated with dignity, as a person whose ideas and views are worth considering, not as troublesome obstacles to the efficient working of the mine site.

Respect for Local Values and Knowledge

Views about where to mine or not mine: As one man from Cajamarca said, "they should respect the decisions of the communities. Accept that there are some places where they can carry out mining, then fine with support from the communities, but accept also that there are some places where mining cannot be carried out, they have to respect the communities" (Yanacocha Narrative 6).

Views about what constitutes "development" or "a good life" (buen vivir):   The narratives reveal that communities are overwhelmingly concerned about the contamination and destruction of water sources which affects their ability to continue farming. There is no sense—from either the Yanacocha or Tintaya narratives—that rural people are willing to abandon their rural livelihoods. Hardly anyone mentioned the improved infrastructure (such as new roads, education or even health) which are listed as the benefits of developments in company facts sheets and websites (Newmont n.d.4) [38]. Access to land and water is vital for them to maintain a rural livelihood. They do not want to be seen as poor, deprived, and lacking but as rural people who wish to continue making a living from the land which is very much part of their cultural identity.

> "…they come and install supermarkets, they come and impose new forms of living that completely contradict with the one that we have" (Yanacocha Narrative 1).

> "[the community] is so divided …our traditions are no longer there, our culture, everything is breaking apart, and the mining companies should not do that" (Tintaya Narrative 3).

For NGOs, respect also extends to the recognition of rural Andean communities as Indigenous communities:

> "…First, the companies should be forced to know the rights of indigenous communities and apply international standards that expect them to consult the communities even if the state does not respect those rights …I think companies should prioritize the rights community have to be informed, to be consulted and to give their consent. If a company does not make sure that this process is uphold and insists to carry on, then they have to accept that together with the state they are violating the rights of indigenous people" (Interview environmental lawyer).

2.7.3   DIALOGUE

There were some references to direct interaction with community engagement personnel but it seems, from the narratives, that interaction with the "company" comes in a variety of guises: with engineers, managers, police, the Public Prosecution Office, and sometimes community engagement people. Not only do people want a chance to speak directly to those who make decisions that affect their lives, but they also need to know who those people are.

> "It would be good for the private companies to open the door to us, for us to speak about how we can live with the mine but instead of that we find the doors closed, we do not have access to dialogue, to speak …that is why we are demanding that the [government] changes their current behavior that is acting in favor of the mining company , not in favor of the communities …the communities that have the impacts …It is not good that we find ourselves marginalised by the mining companies, we are being deprived of our rights by the mining companies" (Tintaya Narrative 1).

Effective Dialogue Should be Inclusive

The message from these narratives is that an effective dialogue has to encompass every section of the community that is impacted by mining.

> "The company should speak, should come and dialogue but they do not come, they do not talk to women. We want to learn. I belong to a mother's club and we are worried about the children …but they never been to the club to talk to us" (Tintaya Narrative 2).

> "in none of the studies that they have carried out …do they consider the *Ronda Campesina* as a relevant actor. It is as if for them they (the Ronda) does not exist …if you are their ally then they apply the politics of a good neighbor, social responsibility, etc., but in the case of the Ronda …[it] is an enemy and so with their enemies they apply very explicit security politics in which the enemy has to be identified, has to be followed, has to be neutralised …they use the word neutralise and so they work with two scripts the social responsible one is applied to those that can be …their allies, those that can defend them and the security politics is applied against all those that are not aligned with their ways …that oppose the general interest of the company and their interest in Cajamarca is expansion …the company has developed many faces …one face is social responsibility for their allies but another face is for repression …In general for community relations they hire people with good intentions anthropologists for example but also many women and these social responsibility people try to do things in one way …according to principles they believe in but they are not aware of all of the other dimensions of the company …because they share very little …and in some cases they are surprised when the other [face of the company] comes to the surface" (Yanacocha Narrative 3).

Effective Dialogue Requires a Review of Time Frames

People who have legitimate queries about mining operations do not want to be excluded from public meetings, or presented with lengthy and unintelligible documents which are difficult to obtain and even more difficult to understand. Time is needed to accomplish this, as well as the acceptance that conflict is always evolving:

> "[in order to get social consent corporations] …need to have the capacity to accept that the real time frames need to be respected not to impose the other time frames …I think conflicts are always going to be present—they are always going to be alive, in essence, they are conflicting relationships, what [governments and corporations] …need is the capacity to act in those conflictive relationships and allow the conflicts to carry on evolving" (Interview academic and NGO member).

Trust

There are many references to broken promises in the narratives, and in the issues which are set out above. These broken promises generate a lack of trust and a general scepticism about the honesty of mining company employees (2.5.1); the integrity of mining company employees (see 2.5.5); the ability of the state to protect community interests (2.5.3).

2.8   PRINCIPLES FOR COMMUNITY ENGAGEMENT

The idea of "community engagement" can mean very different things for communities, for people within communities, for corporations, for states, and for NGOs. Most of those interviewed for this research (whether they wanted to benefit from mining or not) were skeptical about "dialogues," "roundtables," or other encounters. They, not surprisingly, believe that government and companies have an instrumental rationale with which they wish to "restore public credibility, diffuse conflicts, justify decisions and limite future challenges to implementation by 'creating ownership'" (Wesselink and Paavola 2011, p. 2690) [52]. Community principles for engagement, as defined above, are not a means to an end. Rather, they are normative and, furthermore, they clearly align with what is considered "best practice" participation in environmental management which also emphasizes trust, equity, transparency, inclusivity, and co-sharing of knowledge (Reed 2008, pp. 2421–2416) [48].

We have identified a number of shared obstacles and problematic issues in this comparative assessment of relations between these two mining companies and the communities that live alongside them. These obstacles and issues are not unique to these two mine sites (see, for example, Muradian et al. 2003 [43], Arellano-Yanguas 2008 [2], World Bank 2005 [53], Urkidi and Walter 2011 [50], Kemp et al. 2013 [30]). Urkidi and Walter in particular have identified a major challenge in understanding between companies and communities:

> "While Latin American governments and mining companies discuss in terms of revenue and money compensation for externalities, communities are demanding democracy, bottom-up decision making, and recognition of the links between culture and environment. The idea of "good living" (*buen vivir*) that different social actors in Latin America are embracing summarizes a critique of unfair development models and a new approach for thinking about wellbeing" (2011, p. 693 [50]).

Increased trust, transparency, and improved dialogue are key words for these and other reports about community engagement in Peru. To mitigate the issues, an increasing number of consultancies are developing in Lima to support companies with community engagement and "social diagnostics" in order to create the social license to operate. Many companies (both junior and major corporations), as well as the Government, report that community engagement has improved and social conflict can be avoided by these measures. However, it is clear that despite intentions to listen to community, company representatives generate an impression of a lack of respect for local perceptions and declared needs. This reduces their ability to understand community interests necessary to negotiate long lasting relationships.

2.9   FURTHER REFLECTIONS

The crucial issue is to move beyond recognition of the causes of social conflict to recommendations for action that can generate effective change. How do we give meaning to the guiding principles of increased respect, trust, transparency, and effective dialogue between governments, companies and communities? As Harvey points out, much of the literature which recommends improvement in the community engagement sector of mining companies "frequently reads like material we would expect to find emanating from the development sector, rather than tailored advice for the extractive industry" (2013) [25].

In a previous publication, two authors of this report have also argued that recommendations for improvement "need to be anchored in the pragmatic reality of ordinary life for them to have real meaning and value beyond the level of abstract debate" (Armstrong, Baillie, and Cumming-Potvin 2014, p. 92) [4]. Sub-contracting the task of improvement to other companies or agencies is not the solution. This does not require behavioral changes on the part of those in the extractive industry, that is, it does not require them to engage with guiding principles for change.

Bruce Harvey, who has had many years of experience in this industry as Rio Tinto's Global community engagement representative, recommends that employees should participate in local induction courses which should be:

> "crafted and delivered by professional educators with the active involvement of local people. They should be tailored to local circumstance, not framed in universalisms and theoretical abstraction. Based on a desire to instil comprehension more than compliance, local induction should provide local historical and contemporary context and a 'safe' environment for employees and community members to discuss difficult issues" (Harvey 2013, p. 4) [25].

These views are mirrored in another recent report on the costs of social conflict in the extractive sector which also underlines the importance of internal incentives to change (Davis and Franks 2014) [19]. These authors declare the need to address the problematic tension between the time frame required to build productive relationships and the time frame of "short term production targets or ambitious construction schedules …and distinct budget lines" (Davis and Franks 2014, p. 9) [19].

The material in this report is "tailored to local circumstance" and provides the kind of "local historical and contemporary context" that should be the basis for knowledge transfer in the pursuit of institutional change.

Our first recommended actions are therefore the following.

1. Develop local induction courses for company personnel, crafted by professionals together with local people and involve narratives or films depicting local people's experiences.

   We also recommend the support of communities to understand company perspectives and develop the capability to voice their needs as *interests*, crucial for equitable and sustainable negotiation processes. We recommend the following action for further continuation of this critical work.

2. Engage with different levels of mining industry representatives who work or have worked in Peru, and with local government representatives and NGOs, to study their approaches and strategies to *listen* to communities and their response to what they *hear*. This will aim to ultimately facilitate the development of a mutual understanding of company–community *interests*.

   It is imperative that the various researchers and practitioners working to enhance fair and equitable community engagement processes in Peruvian mining contexts collaborate and act together to address these issues. Hence, our final recommended action is:

3. Create a network of actors involved in the support of community engagement processes in Peru.

2.10 REFERENCES

[1] Arana Zegarra, M. (2004). *Impacts of Minera Yanacocha's activities on water resources and the affirmation of citizen rights: The Quilish crisis 2004.* Cajamarca, Grufido. (Translated by Fabiana Li.) 38

[2] Arellano-Yanguas, J. (2008). *A Thoroughly Modern Resource Curse? The New Natural Resource Policy Agenda and the Mining Revival in Peru.* Brighton, Institute of Development Studies. http://r4d.dfid.gov.uk/PDF/Outputs/FutureState/wp300.pdf 45

[3] Arellano-Yanguas, J. (2010). Local politics, conflict and development in Peruvian mining regions. Ph.D. Thesis, University of Sussex.

[4] Armstrong, R., Baillie, C., and Cumming-Potvin, W. (2014). *Mining and Communities: Understanding the Context of Engineering Practice.* Morgan & Claypool Press. DOI: 10.2200/s00564ed1v01y201401ets021. 46

[5] Bebbington, A. (Ed.) (2011). *Social Conflict, Economic Development and the Extractive Industry: Evidence from South America.* Taylor & Francis. DOI: 10.4324/9780203639030.

[6] Bebbington, A. (Ed.) (2012). *Social Conflict, Economic Development and the Extractive Industry.* Routledge. DOI: 10.4324/9780203639030.

[7] Bebbington, A. J. and Bury, J. T. (2009). Institutional challenges for mining and sustainability in Peru. *Proc. of the National Academy of Sciences*, 106(41):17296–17301. DOI: 10.1073/pnas.0906057106. 9

[8] Bebbington, A., Hinojosa, L., Bebbington, D. H., Burneo, M. L., and Warnaars, X. (2008a). Contention and ambiguity: Mining and the possibilities of development. *Development and Change*, 39(6):887–914. DOI: 10.1111/j.1467-7660.2008.00517.x. 9

[9] Bebbington, A., Humphreys Bebbington, D., Bury, J., Lingan, J., Muñoz, J. P., and Scurrah, M. (2008b). Mining and social movements: Struggles over livelihood and rural territorial development in the Andes. *World Development*, 36(12):2888–2905. DOI: 10.1016/j.worlddev.2007.11.016.

[10] Bebbington, A. and Williams, M. (2008). Water and mining conflicts in Peru. *Mountain Research and Development*, 28(3):190–195. DOI: 10.1659/mrd.1039.

[11] Braaten, R. H. (2014). Land rights and community cooperation: Public goods experiments from Peru. *World Development*, 61:127–141. DOI: 10.1016/j.worlddev.2014.04.002. 9

[12] Brush, S. (1976). Man's use of an Andean ecosystem. *Human Ecology*, 4(2):147–166. DOI: 10.1007/bf01531218.

[13] Bury, J. (2004). Livelihoods in transition: Transnational gold mining operations and local change in Cajamarca, Peru. *The Geographical Journal*, 170(1):78–91. DOI: 10.1111/j.0016-7398.2004.05042.x. 9, 10

[14] Bury, J. (2005). Mining mountains: Neoliberalism, land tenure, livelihoods, and the new Peruvian mining industry in Cajamarca. *Environment and Planning A*, 37(2):221–239. DOI: 10.1068/a371.

[15] Bury, J. T. (2007). Livelihoods, mining and peasant protests in the Peruvian Andes. *Journal of Latin American Geography*, 1(1):1–19. DOI: 10.1353/lag.2007.0018. 9

[16] Bury, J. T. (2007b). Mining migrants: Transnational mining and migration patterns in the Peruvian Andes. *The Professional Geographer*, 59(3):378–389. DOI: 10.1111/j.1467-9272.2007.00620.x.

[17] Crabtree, J. and Crabtree-Condor, I. (2012). The politics of extractive industries in the Central Andes. In A. Bebbington (Ed.), *Social Conflict, Economic Development and the Extractive Industry*, pages 46–64, Routledge. 8

[18] Cusicanqui, S. (1993). Anthropology and society in the Andes: Themes and issues. *Critique of Anthropology*, 13(1):77–96. DOI: 10.1177/0308275x9301300104. 9

[19] Davis, R. and Franks, D. (2014) Costs of company-community conflict in the extractive sector. *Report of the CSR Initiative at the Harvard Kennedy School.* http://www.hks.harvard.edu/m-rcbg/CSRI/research/Costs%20of%20Conflict_Davis%20%20Franks.pdf 46

[20] De Echave, J. (2008). *Diez Años de Mineria en el Perú.* Lima, Cooperaccion. 8

[21] Deere, C. (1990). *Household and Class Relations: Peasants and Landlords in Northern Peru.* University of California Press. 10

[22] Dell, M. (2010). The persistent effects of Peru's mining *mita. Econometrica,* 78(6):1863–1903. DOI: 10.2139/ssrn.1596425. 8

[23] Gifford, B. and Kestler, A. (2008). Toward a theory of local legitimacy by MNEs in developing nations: Newmont mining and health sustainable development in Peru. *Journal of International Management,* 14(4):340–352. DOI: 10.1016/j.intman.2007.09.005.

[24] Gutiérrez, R. and Jones, A. (2004). Corporate social responsibility in Latin America: An overview of its characteristics and effects on local communities, *Contreras, ME,* pages 151–188.

[25] Harvey, B. (2013). Social development will not deliver social licence to operate for the extractive sector. *The Extractive Industries in Society.* http://dx.doi.org/10.1016/j.exis.2013.11.001 DOI: 10.1016/j.exis.2013.11.001. 46

[26] ICMM International Council of Mining and Minerals. (2013). *Community Development Toolkit.* http://www.icmm.com/community-development-toolkit 8, 9

[27] IIED International Institute for Environment and Development. (2002). *Breaking New Ground: Mining, Minerals and Sustainable Development.* http://www.iied.org/mmsd-final-report

[28] Kamphuis, C. (2011). Foreign investment and the privatization of coercion: A case study of the forza security company in Peru. *The Brooklyn Journal of International Law,* 37:529.

[29] Kay, C. (2007). Achievements and contradictions of the Peruvian agrarian reform. *The Journal of Development Studies,* 18(2):141–170. DOI: 10.1080/00220388208421824.

[30] Kemp, D., Owen J., Cervantes, M., and Benavides Rueda, J. (2013). *Listening to the City of Cajamarca: A Study Commissioned by Minera Yanacocha.* Brisbane, Centre for Social Responsibility in Mining, University of Queensland. https://www.csrm.uq.edu.au/publications/listening-to-the-city-of-cajamarca-a-study-commissioned-by-minera-yanacocha-final-report 11, 45

[31] Levit, S. (2013). *Glencore Xstrata's Espinar Province Mines: Cumulative Impacts to Human Health and the Environment*. Report prepared for Oxfam America. Centre for Science in Public Participation. http://www.csp2.org/files/reports/Cumulative%20Impacts%20of%20Espinar%20Province%20Mines%20-%20Levit%20CSP2%203Jul13.pdf

[32] Li, F. (2009). Documenting accountability: Environmental impact assessment in a Peruvian mining project. *PoLAR: Political and Legal Anthropology Review*, 32(2):218–236. DOI: 10.1111/j.1555-2934.2009.01042.x.

[33] Laplante, L. J. and Spears, S. A. (2008). Out of the conflict zone: The case for community consent processes in the extractive sector. *Yale Human Rights and Development Law Journal*, 11:69.

[34] Lostarnau, C., Oyarzún, J., Maturana, H., Soto, G., Señoret, M., Soto, M., and Oyarzún, R. (2011). Stakeholder participation within the public environmental system in Chile: Major gaps between theory and practice. *Journal of Environmental Management*, 92(10):2470–2478. DOI: 10.1016/j.jenvman.2011.05.008.

[35] Newmont Mining (n.d.1). *Conga Project Fact Sheet: Citizen Participation.* http://www.newmont.com/sites/default/files/u87/Citizen%20Participation%20Fact%20Sheet%2004%2015%2012%20Final.pdf

[36] Newmont Mining (n.d.2). *Conga Project Fact Sheet: Social Development.* http://www.newmont.com/sites/default/files/u87/Conga%20Project%20Social%20Development%20Fact%20Sheet-010512.pdf

[37] Newmont Mining (n.d.3). *Conga Project Update.* http://www.newmont.com/sites/default/files/u110/Conga%20Update%20Fact%20Sheet%20Final.pdf

[38] Newmont Mining (n.d.4). *Chaupe Family Land Dispute.* http://www.newmont.com/node/5047 43

[39] Newmont Mining (2011). *Conga Project Fact Sheet: Environmental Impact Assessment.* http://www.newmont.com/sites/default/files/u110/Conga%20Update%20Fact%20Sheet%20Final.pdf

[40] Newmont Mining (2013). *Conga Fact Sheet.* http://www.newmont.com/node/4937

[41] Martinez-Alier, J. (2001). Mining conflicts, environmental justice, and valuation. *Journal of Hazardous Materials*, 86:153–170. DOI: 10.1016/s0304-3894(01)00252-7.

[42] Munoz, I., Paredes, M., and Thorp, R. (2007). Group inequalities and the nature and power of collective action: Case studies from Peru. *World Development*, 35(11):1929–1946. DOI: 10.1016/j.worlddev.2007.01.002. 10, 11

[43] Muradian, R., Martinez-Alier, J., and Correa, H. (2003). International capital versus local population: The environmental conflict of the Tambogrande mining project, Peru. *Society and Natural Resources*, 16(9):775–792. DOI: 10.1080/08941920309166. 45

[44] Mutti, D., Yakovleva, N., Vazquez-Brust, D., and Di Marco, M. H. (2012). Corporate social responsibility in the mining industry: Perspectives from stakeholder groups in Argentina. *Resources Policy*, 37(2):212–222. DOI: 10.1016/j.resourpol.2011.05.001.

[45] Orian, E. (2008). The transfer of environmental technology as a tool for empowering communities in conflict; the case of participatory water monitoring in Cajamarca, Peru. Dissertation submitted for Masters of Environmental Science Policy and Management, University of Manchester.

[46] Oxfam (n.d.). *Impacts of Mining*. https://www.oxfam.org.au/explore/mining/impacts-of-mining/

[47] Oxfam (2003). Mining Ombudsman Report tintaya case study. http://resources.oxfam.org.au/pages/view.php?ref=93&search=tintaya&order_by=relevance&sort=DESC&offset=0&archive=0&k=

[48] Reed, M. (2008). Stakeholder participation for environmental management: A literature review. *Biological Conservation*, 141:2417–2431. DOI: 10.1016/j.biocon.2008.07.014. 45

[49] Triscitti, F. (2013). Mining, development and corporate-community conflicts in Peru. *Community Development Journal*, 48(3):437–450. DOI: 10.1093/cdj/bst024.

[50] Urkidi, L. and Walter, M. (2011). Dimensions of environmental justice in anti-gold mining movements in Latin America. *Geoforum*, 42(6):683–695. DOI: 10.1016/j.geoforum.2011.06.003. 45

[51] Urkidi, L. (2010). A glocal environmental movement against gold mining: Pascua–Lama in Chile. *Ecological Economics*, 70(2):219–227. DOI: 10.1016/j.ecolecon.2010.05.004.

[52] Wesselink, A. and Paavola, J. (2011). Rationales for public participation in environmental policy and governance: Practitioners perspectives. *Environment and Planning A*, 43:2688–2704. DOI: 10.1068/a44161. 45

[53] World Bank (2005). *Wealth and Sustainability: The Environmental and Social Dimensions of the Mining Sector in Peru*. Washington, DC, World Bank. http://documents.worldbank.org/curated/en/2005/12/7041000/wealth-sustainability-environmental-social-dimensions-mining-sector-peru-vol-2--2-main-report 8, 45

CHAPTER 3

The Ineffectiveness of Human Rights Protection Mechanisms for Communities Affected by Mining: A Case Study of Minas Conga in Cajamarca, Peru

Jordan Aitken

3.1 INTRODUCTION

The[1] impact of widespread community opposition to major mining projects in Peru has been so significant that influential industry and government figures agree that future mining developments are heavily reliant on how "the wave of social conflict and mining opposition is managed" (De la Flor 2014 [10]).[2]

This sentiment reflects a trend observed across the world, in which society expects corporate actors in the extractive sector to develop and implement increasingly extensive and sophisticated corporate social responsibility (or CSR) policies and practices. Articulated in its most basic form, CSR is the notion that businesses should actively consider and address the social and environmental (in addition to the economic) implications of their operations.

Historically, companies in the extractive sector have implemented CSR policies in an inconsistent manner. Some scholars believe that these practices are adopted merely as "a risk management exercise" used by mining companies to manage social conflict. In contrast, others see CSR and companies' commitment to sustainable mining practices as an example of new revolutionary mining practices (Harvey 2014 [13]; Hilson 2012 [14]).

Regardless of the motive, attaining social acceptance has become an increasingly important factor for all mining projects. In recognition of this requirement, governments and busi-

---

[1]The views expressed in this chapter are the views of the author only and do not represent or reflect in any way the view of any other organization with which author is associated.

[2]This chapter was drafted in 2016, as the author's honor's thesis while completing a B. Eng. at the University of Western Australia, and was edited for publication in 2020.

Table 3.1: Acronyms

Acronym	Description
CLSHRC	Columbia Law School Human Rights Clinic
CSRM	Centre for Social Responsibility in Mining
ICCPR	International Covenant of Civil and Political Rights
ICESCR	International Covenant of Economic, Social, and Cultural Rights
IM4DC	International Mining for Development Centre
LAMMP	Latin American Mine Monitoring Program
NGO	Non-Government Organization

nesses alike have made public commitments to ensure that communities situated near to proposed mine sites are not adversely impacted by proposed operations (Esteves 2008 [12]; ICMM 2013 [75]). Among these are commitments to human rights and environmental protection, community consultation, and ensuring the right to peaceful mobilzation.

Yet, obtaining the social license to operate remains one of the greatest challenges for projects based in Latin America. The concentration of mineral resource deposits in rural areas is a significant factor in this regard. Increased foreign mining investments in Andean countries have coincided with the disruption to the livelihoods of rural communities (CSRM 2013 [36]; Bebbington et al. 2008, p. iv [2]). This has meant that extractive projects across the region frequently face opposition from affected communities.

Equally, community concerns are often treated as subservient to the development of the mining project by governments, owing to many Latin American countries' significant economic reliance upon the extractive sector (CLSHRC 2015 [25])—this dynamic has led, at times, to the repression of community mobilizations. It is therefore not surprising that widespread social conflict has become a relatively common outcome of extractive projects in Latin America (Bebbington et al. 2008 [2]; Defensoría del Pueblo 2012 [27]).

Mining activities in Peru—one of the world's largest-scale producers of base and precious metals—in particular, have drawn international attention. Peru's economic prosperity is closely linked to its commodity exports. It has enjoyed a stable political and macroeconomic environment since 2000, owing in part to prudent economic reforms. This relative stability, coupled with the abundant and high-grade mineral deposits across the country, helped to attract significant foreign investment that underpinned strong economic growth during this period (most significantly from 2002–2013). As a result, the Peruvian population enjoyed significant economic benefits: a substantial reduction of poverty (26% between 2005 and 2014) and high volumes of revenue ($27.4 billion at its peak) (The Economist, 2016) [67].

However, during this same period, NGOs and human rights groups have consistently expressed concerns over a cycle of violence in Peru perpetrated by the mining sector, including state security forces' repression of peaceful mobilization against irresponsible mining practices (Earth Rights International 2014b [30]; Human Rights Watch 2015 [34]). A case study of the Minas Conga mine in Cajamarca in Peru's northern highlands illustrates the heavy-handed policies and practices of the Peruvian government that have become common across the country.

Minas Conga, which was majority owned by American gold mining giant Newmont, was due to become one of the largest mining projects in South America. The company's proposed plan to develop the mine included draining four lakes. These lakes continue to be critical water sources for the expanded surrounding region and embody great spiritual significance to the local community, which mobilized to express their opposition to the mining proposal. In response to the community's opposition, the operator and the Peruvian government acting in partnership established an environment of repression and intimidation against those who opposed the project (IM4DC 2014 [24]; CSRM 2013 [36]). This is despite a range of commitments, by both the state and the operating company, to protect and promote human rights standards.

This chapter seeks to demonstrate the relative ineffectiveness of the various government and business commitments to protect and promote human rights when such rights are obstacles to commercial mining interests—especially for remote communities impacted by mining project. This chapter also analyzes the various ways in which the State and company failed to ensure that communities adjacent to the proposed mine site were not adversely impacted. Further, it exposes the significant repurcussions for projects that fail to adequately engage with impacted communities and consequently do not obtain the social licence to operate.

3.2   LITERATURE REVIEW

3.2.1   A HISTORY OF MINING AND SOCIAL CONFLICT IN PERU

Mining activities are intertwined with Peru's history and continue to be a central tenet of its economy. In 2015, Peru was the world's third largest producer of silver, tin, zinc, and copper, and fifth largest producer of gold (United States Geological Survey 2015) [73]. In fact, mining accounts for approximately 14% of Peru's gross domestic product, and approximately 60% of its exports (de la Flor 2014) [10]. However, Peru's mining sector also has strong links to slavery, harsh labor and economic inequality. As a result, large sections of Peruvian society harbour deepseated resentment toward the industry (World Bank 2005 p. 17) [41].

Peru's heavy economic reliance upon mining exports was further consolidated by market reforms implemented by the government in the early 1990s (Crabtree and Crabtree-Condor, 2012 p. 50) [7]. These reforms, which included generous tax incentives for large multinational companies, created a more stable and attractive environment for foreign investors (Arellano-Yanguas 2012) [1]. On the back of these reforms and driven by increased output from the extractive sector, the economic conditions in Peru improved significantly: government revenue increased, inflation slowed, and general economic stability was established (IM4DC 2014 p. 9) [24].

While mining activities have contributed to the general health of Peru's economy, they have disproportionately affected its rural communities. This is because many of the existing and proposed mining projects in Peru are situated in remote locations in the Andean highlands and rainforests (Bebbington et al. 2008) [2]—in close proximity to Peru's indigenous communities. In fact, by 2010, over half of Peru's peasant communities lived in regions impacted by mining activities (IM4DC 2014 p. 8) [24]. These communities have traditionally relied upon subsistence farming and other agricultural-based activities to support their livelihoods (Bury and Kolff 2002) [6], and are generally not well serviced by basic public services. As a result, access to reliable clean water sources is integral to the sustenance of their livelihood (Braaten 2014) [5]. Furthermore, indigenous Peruvians also have a strong spiritual connection with the land: especially the mountains, the lakes and the forest (IM4DC 2014) [24].

As mining activities in Peru have modernized, large-scale open-pit mining has emerged as the preferred mining technique. However, the environmental footprints of modern open pit mines have become exponentially greater than those of alternative, smaller-scale mining methods (Triscritti 2013) [23]. Indeed, there are numerous examples where open pit mines have impacted natural water sources and adversely affected the traditional livelihoods of adjacent communities (Bebbington and Williams 2009) [4].

These negative environmental impacts are frequently cited by opponents to mining projects in Peru (Bebbington et al. 2008) [2], as well as the mining companies' failure to adequately consult impacted communities in order to obtain free and informed consent prior to commencing operations (Triscritti 2013 p. 440) [23].

The unavailability and ineffectiveness of formal channels of complaint (Bebbington 2012) [3] and the perception that favorable treatment is afforded to mining companies by the government have meant that affected communities have often resorted to using peaceful protest to express their opposition (CSRM 2013 [36]; Bebbington et al. 200 [2]). For instance, substantial anti-mining protests erupted at Rio Blanco (2005), Rio Tinto's copper project in Lambayeque Province (2008), and the Yanacoha mine in Cajamarca (2008). Community led protest efforts are increasingly supported by NGOs and other bodies that recognize the power imbalance between the communities and the State (Crabtree and Crabtree-Condor 2012) [7]. As a result, anti-mining mobilizations in Peru now reach a broader audience and present a greater obstacle to major projects.

Protest is not a new negotiating tool for Latin American communities. However, since the presidency of Alberto Fujimori in the 1990s, Peruvian governments have increasingly resorted to the use of force and intimidation to halt peaceful community mobilizations (The Economist 2016 [67]; Internacional de Resistentes a la Guerra 2015 [56]; Bebbington et al. 2008 [2]). Community opposition is often stifled through the criminalization of protest (Triscritti 2012) [72] and the deployment of disproportionate police presences in areas adjacent to the protest site (Holden and Jacobson 2007 p. 478) [15]. Consistent with this trend, recorded social conflicts rose from 73–215 per month in just five years (between January 2006 and September 2011).

During this period, there were 195 deaths and 2312 injuries directly related to protests in Peru (Defensoría del Pueblo 2012) [27].

Despite international criticism from NGOs, media organizations, and human rights bodies (Human Rights Watch 2016) [33], repressive practices remain a common response to anti-mining protests in Peru.

3.2.2 MINAS CONGA

Minera Yanacocha is a joint venture company majority owned by Newmont Mining Ltd. It launched the Minas Conga project in 2004—a proposed gold mine situated in the Cajamarca district in Peru's northern highlights. The project was designed as an expansion of the company's Yanacocha mine, which commenced operation in 1993. The planned $4.8 billion investment was set to become the largest gold mining operation in Latin America and Peru's largest single foreign investment (Reuters 2011) [66].

However, Conga experienced strong opposition from the adjacent communities since the project's inception. Various opposition movements have declared that the social and environmental impacts of the proposed mine are unviable (Defensoría del Pueblo 2010) [26]. The opposition has concentrated on the mine's plans to drain four natural lakes and replace them with four constructed reservoirs (Triscritti 2013) [23]. These four lakes are the primary water source for the surrounding region, and to many local residents, are indispensable to their livelihood.

"[protest] is about the life of the actual communities, because, we realized that the mining company had located itself where the rivers are born, the same rivers that give water for agriculture, livestock, for human consumption of the communities…" (P_1).

The legitimacy of these concerns was acknowledged by Peru's Environment Minister, Mr Ricardo Giesecke, in 2011. Mr Giesecke expressed that *"the Conga project will transform in a significant and irreversible manner the river basins"* (Poole and Rénique n.d.) [64] and claimed that draining the lakes would be equivalent to *"dynamiting the glaciers in the Andes"* (Reuters 2011) [66].

Equally, opposition to Minas Conga has also been fueled by the community's distrust of Minera Yanacocha. Throughout 20 years operating in Cajamarca, the company has been responsible for numerous controversies, including a significant oil spill that contaminated drinking water and food sources, as well as intermittent threats to local water supply, which have disrupted community life (CLSHRC 2015, p. 3) [25]. As a result, the community does not trust Yanacocha to operate without polluting the environment (Triscritti 2013 [23]; CSRM 2013, p. 14 [36]).

Despite the community's concerns and strong opposition, the project's environmental impact assessment was approved by the Peruvian government in 2010. Newmont proceeded with early construction for the project in 2011, commencing work on one of the reservoirs. Around this time, protests against Conga gradually gathered momentum. The protests were supported by regional leaders, including the President of the Cajamarca Region, Mr. Gregorio Santos. By

September 2011, community leaders had declared that Conga would not receive a "social license" under any circumstances (Triscritti 2013, p. 442) [23]. At around the same time, protests intensified, culminating in hundreds of community members congregating at the mine site. In light of the mounting pressure, Newmont halted operations at Conga in November 2011. Shortly, thereafter, the Peruvian government mobilized its security forces to attempt to suppress the movement, which triggered a cycle of conflict in Cajamarca and the surrounding area (Triscritti 2013) [23]. Table 3.2 is a timeline summarizing the key events in the Conga's lifespan (until 2012).

3.3   METHODOLOGY

3.3.1   RESEARCH OBJECTIVE

This study is a qualitative examination of the use of mobilization by communities impacted by mining operations in Peru. It also examines the responses of business and government, analyzed through the lens of their respective human rights commitments. Specifically, the study focuses on the rights to peaceful assembly, freedom of association and freedom of expression—all of which are codified in the *International Covenant of Civil and Political Rights* (ICCPR), to which Peru is a party.

It explores how the absence of grievance mechanisms and community representation in the national government (De la Flor 2014) [10] transformed the exercise of these rights into critical tools for the defense of communities directly affected by mining projects. It also explores the idea that, at times, these rights (and commitments by states and businesses to respect and uphold them) do not afford vulnerable communities the protection that the rights envisage.

The study provides a platform to project the "community experience," which is regularly underrepresented in mining-related conflict discourse. Identifying the corporate behaviors that led to or resulted in negative outcomes during the Conga project (in particular, the absence of clear and effective community engagement) could inform the development of improved community liaison strategies. Increased corporate awareness of the complexities of community engagement could also assist the development of mining policies and practices that are more responsive to community grievances and concerns.

3.3.2   DATA COLLECTION

The study used a collection of testimonial and documentary data, compiled from primary and secondary research carried out across Australia, Peru, and the United States. It sources interviews conducted in Peru with members of communities impacted by the Conga project, members of NGOs and human rights activists as part of a project undertaken by IM4DC. These interviews were used with the permission of the interviewers. The author conducted an interview with a senior employee at Newmont, who has an intimate knowledge of Newmont's Peruvian operations. Interviews were open-ended, but guided by a common structure. Due to the transparency

Table 3.2: Timeline of key events at Conga (RESOLVE 2016 [39]; IM4DC 2014 [24]; Triscritti 2013 [23]; CSRM 2013 [36]; CLSHRC 2015 [25]; Earth Rights International 2014a [29]; Newmont 2016a [58]; Newmont 2015 [60]; Newmont 2014a [61]) (*Continues.*)

Date	Event
2004	Newmont announced the project to go ahead as an extension of Yanacocha mine.
Oct. 27, 2010	Environmental Impact Assessment approved following a review and community consultation.
Early 2011	Construction of the Chailhuagon Reservoir commences.
May 25, 2011	Jamie Chaupe Lozano files a claim against Yanacocha for invading his family's property
Aug. 10, 2011	Jamie Chaupe Lozano complains to Peru's Ombudsman about Yanacocha's attempted forcible evictions.
Aug. 11, 2011	Public Prosecutor in Celendin closes Jamie Chaupe's complaing of May 25, 2011. Minera Yanacocha files a criminal complaint against the Chaupe family.
Oct., 2011	Anti-mining activists expressed concerns about perceived water impacts.
Nov., 2011	Protesters begin to asemble in hundreds at the Minas Conga concession site.
Nov. 24, 2011	A general strike is declared throughout Cajamarca as protesters assembled at the Conga concession area. Approximately 30 police officers ordered protesters to disperse. This was followed by firing tear gas, rubber bullets, and live ammunition. 24 protesters were injured.
Nov. 30, 2011	Newmont suspends construction at the concession site, citing concerns for the safety of employees and community members.
Dec. 5, 2011	Peruvian President Ollanta Humala declares a regional State of Emergency in Cajamarca for 60 days.
Dec. 22, 2011	Celendin Court grants a request to evict Chaupe family from their property. Minera Yanacocha does not act on the court order due to the political climate.
Feb. 9, 2012	Thousand of protesters march from Cajamarca to Lima (taking 9 days) for the National March in Defence of Water. The protesters presented Congress with legislative proposal to ban mining in headwaters.
Feb. 27, 2012	Peruvian government hires consultants from Spain and Portugal to undertake a 40-day review of the existing Environmental Impact Assessment.
April, 2012	The panel hired to review EIA finds that it is acceptable.
May, 2012	A general strike is declared again.
May 31, 2012	Government forces knock over cooking pots comprising food for protesters at the concession site. They fire tear gas at protesters, 4 people injured, 8 arrested.
June 2, 2012	Two human-rights lawyers are arrested.

Table 3.2: (*Continued.*) Timeline of key events at Conga (RESOLVE 2016 [39]; IM4DC 2014 [24]; Triscritti 2013 [23]; CSRM 2013 [36]; CLSHRC 2015 [25]; Earth Rights International 2014a [29]; Newmont 2016a [58]; Newmont 2015 [60]; Newmont 2014a [61]) (*Continues.*)

July 3, 2012	Regional State of Emergency declared for the second time. Military and police interrupt protesst in Cajamarca. Police open fire with tear gas and live ammunition on protesters. Five people are killed (including one person shot from a helicopter) and numberous injuries are sustained.
July 2, 2012	More than 100 farmers reported for criminal offence investigations for opposing the mine.
Aug., 2012	All works at Conga are suspended.
Oct. 29, 2012	Chaupe family are convicted of "aggravated usurpation" in a Celendin Court. The sentence is appealed.
Jan 25, 2013	Newmont announce new $150 million investment in Minas Conga.
Mar. 13, 2013	Application made to the Inter-American Court of Human Rights for protection measures for Conga protestors.
April 2, 2013	Newmont announce plans to reallocate funds from Conga to other investments.
May, 2013	Construction of the Chailhuagon Reservoir is completed.
June 18, 2013	A new wave of protests threatens as thousands of farmers march on El Perol Lake.
June 19, 2013	Protesters set up camp in the mountains in an attempt to save the lakes.
Aug. 2, 2013	The conviction against the Chaupe family is annulled by the Superior Court of Cajamarca. A re-trial is ordered.
Jan. 17, 2014	Protesters kidnap a mine security worker and vandalize the mine's property.
March 2014	Police security personnel alleged to have forcibly dispersed the protesters conducting a peaceful vigil by firing live ammunition, setting fire to campsites, and burning food, clothing, and equipment.
May 5, 2014	Inter-American Court of Human Rights grants precautionary measures in favor of 46 protesters at Conga.
June 25, 2014	Regional President Gregorio Santos is arrested for opposing central government.
Aug. 5, 2014	Chaupe family members are convicted of aggravated usurpation. The sentence is appealed.
Dec. 17, 2014	Superior Court of Cajamarca revokes the sentence against the Chaupes. It further orders cessation of the preventive eviction and provisional supervision of the Chaupe property by Minera Yanacocha.

Table 3.2: (*Continued.*) Timeline of key events at Conga (RESOLVE 2016 [39]; IM4DC 2014 [24]; Triscritti 2013 [23]; CSRM 2013 [36]; CLSHRC 2015 [25]; Earth Rights International 2014a [29]; Newmont 2016a [58]; Newmont 2015 [60]; Newmont 2014a [61])

Feb. 5, 2015	Alleged demolition of preliminary construction of a new home for the Chaupe family.
Aug. 6, 2015	Maxima Acuna de Chaupe reports intimidation and death threats.
Nov. 20, 2015	Alleged break-in at the Chaupe family home.
April 18, 2016	Newmond stock market release announces that Conga project is indefinitely suspended, re-categorising it as a mineral reserve.

and honesty of the interviewees, accounts of events have been treated as hard data in this study. The interviews have been supplemented with company and government reports and archival research. The interviews are displayed in the text with a code letter and number to protect the rights of the individuals.

3.3.3 DATA ANALYSIS

The data that formed the basis of the study has been analyzed through the lens of the respective human rights commitments of Peru (as a State party to all of the major international human rights treaties, including the *ICCPR*) and Newmont (under its public commitments to various business and human rights initiatives, including the UN Guiding Principles on Business and Human Rights, 2011). In particular, the analysis considers four key issues: the mechanisms adopted by the company and the government in response to the community moblization; the factors that motivated government and company responses to protest; the extent to which relevant human rights commitments provided effective protections to the community; and how the issues raised in this study reflect the broader experience of Peruvian and Latin American communities impacted by mining projects.

3.3.4 POTENTIAL LIMITATIONS

The study acknowledges the role of bias from all of its subjects. In the context of this study, companies and government are likely to frame their actions favorably, most significantly in the context of their responses to protests. Similarly, community allegations of human rights violations and abuses may be exaggerated. To improve its reliability, the collected interview data has been cross-referenced against independent academic research throughout the chapter.

3.4   STATE AND BUSINESS HUMAN RIGHTS COMMITMENTS

The following section provides an overview of the relevant international human rights instruments and business commitments to respect human rights—including the rights to freedom of peaceful assembly, freedom of association, and freedom of expression. The study in the following section will be discussed in light of these commitments.

3.4.1   THE STATE, HUMAN RIGHTS, AND THE RIGHT TO PROTEST

The protection of human rights has become an integral part of the post World War II international order and enjoys universal support among members of the United Nations. Our contemporary conception of human rights was articulated in the Universal Declaration of Human Rights of 1948, which has informed the subsequent development of international treaties that impose legally binding obligations upon their states parties. Of these treaties, the most universally ratified (and arguably most significant) is the *International Covenenant of Civil and Political Rights*, or the *ICCPR*. Peru is a state party to the *ICCPR*, as well as the *International Covenant of Economic Social and Cultural Rights* (*ICESCR*) and numerous additional human rights treaties.

The *ICCPR* includes a range of what are generally considered to be "core" human rights obligations. This includes the right to peaceful assembly, the right to take part in public affairs, the freedom of expression, and freedom of association—which collectively comprise the framework upon which the state ensures that its duty to facilitate peaceful protests is fulfilled (Human Rights Council 2013). Observation of this duty recognizes the importance of peaceful assembly as a platform for change and the advancement of human rights (O'Flaherty 2014) [19].

Consistent with this human rights framework, States may impose certain restrictions upon peaceful demonstrations, but only those that are necessary in the interests of national security, public safety, or public order. In such circumstances, it is incumbent on the State to demonstrate the restrictions are necessary, proportionate, non-discriminatory, and do not impair the democratic functioning of society. State governments must also pursue the least intrusive means of achieving its objective, and police must remain impartial during the conduct of protest. Further, the dispersal of peaceful assemblies and forceful intervention must be the last resort, following attempts to de-escalate violence peacefully (Human Rights Council 2013; Martin 1986 [18]). Finally, no individual or group should be criminalized or subjected to threats or acts of violence because of their involvement in peaceful protest (Human Rights Council 2013).

3.4.2   MINING INDUSTRY'S HUMAN RIGHTS COMMITMENTS

The protection of human rights is an obligation or duty that is owed by States, pursuant to international treaties as well as customary international law. Businesses (which are not subjects of international legal obligations) do not *owe* human rights obligations to individuals or communities. Nonetheless, recently there has been a push by businesses to recognize the corporate responsibility to respect human rights at both social and environmental levels (Ruggie 2007) [20]. This reflects a growing expectation that businesses should respect and uphold human rights standards in the conduct of their operations. In the mining industry, commitments to uphold human rights, including the right to self-determination and the various rights that protect peaceful protest, are vital to the maintenance of responsible and sustainable practices, particularly when vulnerable populations are impacted by mining policies and activities (ICMM 2015) [55].

Three voluntary documents, which have recieved broad international recognition, encapsulate the business commitments most relevant to human rights protection in the mining industry:

- *The UN Guiding Principles on Business on Human Rights, 2011* (the Guiding Principles)

- *The Voluntary Principles on Security and Human Rights, 2000* (the Voluntary Principles)

- *The UN Global Compact, 2000* (the Global Compact)

The UN Guiding Principles on Business and Human Rights outlines a set of guidelines and undertakings for companies (and States) aimed at preventing, addressing, and remedying human rights abuses committed in the course of business operations. Significantly, this includes businesses' responsibility to avoid infringing on the human rights of others and the need to address any adverse human rights impacts with which they are involved (Guiding Principles 2011, Principle 11) [24]. Businesses are expected to seek to prevent or mitigate adverse human rights impacts directly (or indirectly) related to their operations through communication, cooperation, and remediation processes (Guiding Principles 2011, Principle 13) [24].

The Voluntary Principles on Security and Human Rights were developed to address issues in the extractive sector that relate to security arrangments for mining operations. Specifically, the Voluntary Principles aim to ensure the protection of, and respect for, human rights by guiding companies to undertake comprehensive human rights risk assessments in the engagement of security providers for project sites (Voluntary Principles 2000) [45]. Key undertakings include human rights risk assessments, regular consultation with host governments about the impact of security arrangements, maintenance of the rule of law and safeguarding human rights, as well as investigating all credible allegations of human rights abuses (Voluntary Principles 2000) [45]. The Voluntary Principles aim to reduce human rights abuses caused by security arrangements, and in the event that abuses do occur, to mitigate their impact.

The UN Global Compact, the world's largest corporate sustainability initiative, encourages businesses worldwide to adopt sustainable and socially responsible policies. It specifically refers to the duty of businesses to ensure that they are not complicit in human rights abuses (UN Global Compact 2000, Principle 2) [46].

3.4.3 NEWMONT'S HUMAN RIGHTS COMMITMENTS

Newmont has publicly committed to uphold and respect the human rights of people impacted by their operations (Newmont 2015a) [60]. It was a founding member of the International Council on Mining and Minerals in 2002, formally joined the Voluntary Principles in 2002 (Newmont 2015a, p. 2) [60], is a member of the UN Global Compact, and adopted the Guiding Principles in 2014 (Newmont 2014a) [61].

3.5 ANALYSIS

Both the Peruvian government and Newmont (Minera Yanacocha) have made unequivocal commitments to respect and protect human rights (noting, of course, the fundamental difference in the nature of their respective human rights commitments articulated above). The following section analyzes the extent to which each actor failed to uphold and adhere to their respective commitments. For the purposes of ensuring the analysis is completed chronologically, the study begins by first examining Newmont's role in the conflict at Minas Conga, followed by the Peruvian government (and its security forces).

3.5.1 NEWMONT'S PROTECTION OF COMMUNITY RIGHTS

Lack of Consultation

It has been well documented that the Peruvian government approved the Conga project without Newmont first obtaining the acceptance of the affected community, which had expressed strong opposition well before the project's commencement. According to (N_5), "*there was an absence of meaningful consultation with the impacted community,*" which left its members feeling as though it had been deprived "*the opportunity to consent [to] extractive activities that are going to have a significant impact [on] their life.*" As noted by N_4, such practice is reasonably common in Peru: "*when the government give exploitation rights to the companies they do not inform the communities.*"

Newmont's approach to the Environmental Impact Assessment (EIA) was indicative of its failure to meaningfully engage the community. As an element of the EIA it prepared and delivered for Conga, Newmont controversially proposed the construction of four reservoirs to "replace" the lakes at the site, which it pitched as a sustainable environmental solution (Triscritti 2013) [23].

Newmont made tokenistic efforts to engage the community on the EIA by making the document publicly available to the community. However, the document was over 1,000 pages long and steeped in technical language, making it inaccessible and incomprehensible to the broader community. No efforts were made to explain the environmental implications of the project (Li 2009 p. 232) [17]. Only one community consultation session was held approximately a five-hour's drive from the project site, and during the consultation, each community member's input was restricted to one-minute slots (IM4DC 2014) [24]. Despite all of these factors, the EIA was approved only one month after it was first presented to the community for consideration.

The presentation of the EIA did not afford the community a reasonable opportunity to express its views and concerns. The process appeared to be an attempt to bypass consultation, demonstrating the company's indifference to the concerns and interests of the community. Although Newmont did undertake some initiatives for community participation, including the community information center (Newmont 2012) [47], these were perceived as little more than symbolic (Dolan and Rajak 2016) [11]. Ineffective consultation set the tone for the ensuing community distrust of the company at Conga: *they lied to us... and the welcoming attitude [of the community] began to change* (N_2).

Harassment, Property Damage, and Forced Evictions

The regulatory framework governing the acquisition of resource extraction rights in Peru is complex and time consuming (Delgado and Daracco 2019) [28]. A company must, as a first step, acquire sub-surface mining exploration rights within a defined area (Latin Resources n.d.) [57]. Prior to the commencement of any exploration work, the company is required to complete a preliminary EIA, which must address (among other issues related to the nature of the project) the presence of indigenous communities that would be impacted by the footprint of the proposed site. The company is required to develop a more comprehensive plan that addresses the environmental and social impacts of the project once approval to commence exploration activities has been received. Simultaneously, the company must negotiate the purchase or lease of the surface land from its independent owners.

In Cajamarca, rural communities often collectively own the land they occupy (IUCN 2016) [35]. The general community is responsible for allocating the possessory rights over individual lots to its members. As a result, companies generally need to follow a two-stage process to obtain the rights to undertake its mining activities. First the property is purchased from the general community, which requires a two-thirds approval from the "qualified community members." Subsequently, the company negotiates with individual lot holders for possessory rights to each independent lot (RESOLVE 2016, p. 19) [39].

Negotiations with "possessors" of individual lots have proven one of the most controversial aspects of the Conga mine (CLSHRC 2015) [25]. Community accounts have described these efforts as *the first big problem* (N_2), responsible for *the most serious conflicts* (P_1).

Of particular concern, there is evidence that Newmont's security forces and employees have engaged in acts of intimidation and harassment that have threatened the physical integrity of local families and their properties (CLSHRC 2015) [25]. According to C_3, *if you don't get out willingly and if the owner doesn't accept [the offer], they simply threaten you.*

Forced evictions are inconsistent with international human rights law, in particular the right to adequate housing—unless the free, prior, and informed consent of those being displaced has been sought, and the eviction is approved by a State authority vested with power to grant this decision (OHCHR, 2014) [24]. Even when forced eviction is inevitable, businesses should not actively participate in forcible attempts to remove people from their residence.

Newmont's failure to respect and uphold these rights was drawn into public focus when the company attempted to evict the Chaupe family, subsistence farmers who owned a lot in the Sorocucho rural community (RESOLVE 2016) [39]. The conflict started after Newmont accused the family of occupying land previously sold to Minas Conga by the community (Newmont 2015b) [29]. However, during their first confrontation in 2011, company personnel demanded that the family vacate the premises. Since then, the family, supported by local NGOs, has documented the numerous incidents where police and security personnel have intimidated family members, These incidents ranged from trespassing on the land, to issuing death threats, and destroying their property, including crops (RESOLVE 2016) [39].

In one reported incident, Maxima, the matriarch of the Chaupe family, and her daughter Jilda were brutally beaten by police. Maxima describes the events:

"the DINOES dragged [Jilda] by the hair, they kicked her, and they beat her with the butts of their rifles ... The police grabbed me, three in each arm, and they beat me, they left my ankles black."

Video footage from this incident confirms there was an excessive police presence. Further, medical certificates from five days after the incident corroborate the traumatic injuries sustained by Jilda and Maxima (LAMMP 2014) [37].

The conduct described in these allegations exemplifies Newmont's failure to act consistently with its various commitments to respect human rights, including to not participate in, or contribute to, forcible evictions. This was in fact highlighted by the Inter-American Commission on Human Rights' award of precautionary orders for the Chaupes and 46 other protest leaders in 2014 (Inter-American Commission on Human Rights 2014) [43], in recognition of the ongoing risk to their physical security. Regardless, Newmont continues to assert it merely exercised its lawful rights as a landowner and that police, its security personnel and mine employees have never made forceful attempts to evict or intimidate the Chaupe family (Newmont 2016b) [59].

Private Security Contracts with Public Forces

In Peru, for a short period of time, agreements between police and private enterprise were officially sanctioned by Supreme Decree 004-200 (Mora 2015) [71]. Under these arrangements, police officers could be contracted by companies to provide "additional" services on either a permanent or occasional basis (GRUFIDES 2013, p. 9) [32].

The Voluntary Principles address arrangements of this very nature in the context of mining operations. The Principles stress the necessity for public forces to retain impartiality, in order to safeguard against the risk of perpetuating human rights abuses (Voluntary Principles 2000) [45].

Newmont entered into agreements with police to provide private security for the mine pursuant to Supreme Decree 004-2009. Newmont assessed that the Peruvian government considered such arrangements were important for the protection of the country's assets.

> *"they are recognizing that the police have a critical role to play in not only protecting its people, but also for protecting the assets of the country …include[ing] its resources"* (M).

However, the lack of public transparency of the agreements, complemented by the economic incentives for police under these arrangements (in a region of extreme poverty) raised serious concerns about their ability to remain impartial (GRUFIDES 2013) [32]. The relationship between Newmont and the police at Conga gave the community the perception that the *"police [were] aligning themselves with the interests of the mine"* (N_2), and contributed to the human rights abuses and repression of peaceful protest efforts (Mora 2015) [71]. N_4 notes that protection is only available to those with money:

> *"putting public forces at the service of private interests …is a very discriminatory issue, …the one who receives protection from the States is the one who can pay for it."*

Overall, these security arrangements contributed to the sense that *"the company is here colluding with …the police"* (N_5), and the further powerlessness of the community members (Front Line Defenders 2016, p. 3) [53]: *"there are not authorities …they are most like the enemy for us"* (P_3).

The Peruvian government eventually repealed the laws that permitted such arrangements, owing to consistent and widespread international criticism (GRUFIDES 2013 [32]; Human Rights Watch 2012 [33]). The head of the Prevention of Social Conflicts and Governability of the Ombudsman's Office acknowledged the importance of maintaining a *"full time"* and independent police force (Mora 2015) [71]:

> *"We cannot have a division with one hand protecting private interests and the other defending the rights of the people in general."*

Even despite Peru's recognition of the detrimental effect of the law, Newmont remains supportive of re-establishing the arrangements should it become legal again.

> *"…we are supportive of the Peruvian national police in working with the private sectors and companies, and we will strive to be open and transparent around the agreement if and when it becomes legal again"* (M).

Although the security agreements between Newmont and the Peruvian police were operational for a limited duration only, these arrangements catalysed the community's permanent loss of confidence in the police force's ability to remain impartial in its exercise of public duties. In particular, the excessive use of force by the police to suppress protest during this period blurred the distinction between the company and the State—to the extent that the community believed that Newmont was colluding with the police.

3.6   THE PERUVIAN GOVERNMENT

3.6.1   CONSISTENT POLICE PRESENCE AND "STATE OF EMERGENCY"

Peaceful community protests against Minas Conga grew in scale and intensity throughout 2011 (see the timeline in Table 3.2). The Peruvian government deployed a range of tactics in response that effectively denied the community the full enjoyment of their rights, including those guaranteeing their right to peaceful protest.

First, the escalation in protests was matched by equivalent growth to the regional police presence, particularly in major towns. This sent a very clear signal that the Peruvian government would have zero tolerance for opposition to Minas Conga. According to P_2, on one instance *"the police and the army came in their hundreds possibly thousands and basically filled the whole town,"* while a separate account confirms over 1,200 officers in Celendin in 2013 (AAP 2013) [47]. Civil society joined protesters in denouncing the consolidated police presence, which they described as a disproportionate threat (Front Line Defenders 2014) [31], particularly given the peaceful nature of the mobilizations.

Further, on multiple occasions, public security forces resorted to violence to repress the demonstrations (CLSHRC 2015, p. 17) [25]. Recorded incidents have captured the indiscriminate use of tear gas, brutal beating of protesters, and firing upon demonstrations with rubber and live ammunition (Earth Rights International 2014a [29]; Triscritti 2013 [23]).

The State also deployed the declaration of a "state of emergency" as an additional tactic to suppress the protests. Pursuant to Peru's constitution (and consistent with its human rights commitments), the President may lawfully declare a state of emergency, but only in a very narrow range of circumstances (response to disturbances of the domestic order or national disasters). This limitation is critical, because such declarations have the practical effect of temporarily suspending a range of civil liberties, including the right to personal freedom and security, the inviolability of the home, and freedom of assembly and movement in the territory, while additionally authorizing military patrols (Congreso de la República del Perú 1993, Article 137). It is necessary to note that such declarations are consistent with human rights law during a time of national emergency, to the extent strictly required by the situation, and where the deceleration is not discriminatory (ICCPR Article 4(1)).

The President declared a "state of emergency" on two independent occasions in relation to Conga. On the first occasion, a state of emergency was declared on December 5, 2011, following violent clashes on November 29, 2011, when members of DINOES fired teargas, rubber bullets, and live ammunition at unarmed protesters (Earth Rights International 2014a) [29]. A second state of emergency was declared on July 4, 2012, the day after 3,000 protesters marched upon the town of Celendin. On this occasion, military personnel were deployed to "restore peace and internal order" to the region (El Comercio 2012) [52]. Ensuing clashes between protestors and security forces resulted in 5 civilian deaths and at least 20 civilians were injured with gunshot wounds (IM4DC 2014) [24]. Observers reported indiscriminate firing of tear gas and rubber and live ammunition from police on the ground and in helicopters (Sullivan 2014, p. 134) [22].

"There's videos of this, of police directly shooting into the crowd there is also tear gas and plastic bullets fired directly into the crowd" (P$_2$).

During both states of emergencies, testimonial and documentary evidence has demonstrated that State security forces were responsible for repeated conduct that was inconsistent with Peru's human rights commitments. Reports concerning the first state of emergency (November 2011) suggest that it was deployed as a political vice that aimed to quash the prospect of further mobilizations across the country (Poole and Rénique n.d.) [64], rather than for its mandated purpose, a "last resort mechanism; even despite the President's account that the government had *exhausted all paths to establish dialogue as a point of departure to resolve the conflict democratically*" (The Guardian 2011) [68].

Police activity during both states of emergencies was characterized by targeted and indiscriminate uses of force. During the second state of emergency, prominent protest leader Marco Arana was arrested and subsequently beaten for "organizing meetings": *"in the police station they hit me again, punches in the face, kidneys, insults"* (Al Jazeera 2012) [48].

Considered in the broader Peruvian context, it is clear that declarations of states of emergency have become a frequently utilized tool for the government to deploy state security forces en masse to repress opposition during mining-related conflicts. For instance, on two other separate occasions, civilians were killed in protests against the Espinar mine in Cusco, and the Tintaya mine in 2012 (Front Line Defenders 2014, p. 2) [31]. Additionally, a 60-day state of emergency was declared in May 2015 in Arequipa after protests against the Tia Maria mine resulted in deaths, and a 30-day state of emergency in Apurimac in September followed the death of four protesters opposed to Las Bambas mine (BBC 2015) [49].

3.6.2   CRIMINALIZATION OF PROTEST

The State's efforts to repress protest movements have also adopted non-violent forms. The Peruvian government has consistently used the legal system to attempt to punish protesters, particularly those organizers of anti-mining mobilizations. The government tends to exploit broad definitions of terms in the penal code, such as "public intimidation," "incitement to violence," "terrorism," and "kidnapping," which are open to arbitrary interpretations by the judiciary (Catapa, n.d.) [51]. Efforts to criminalize protest activities in this way have been so effective that as of June 2014, nearly 400 anti-mining protesters and human rights activists faced court proceedings initiated by the company or the public prosecutor (Front Line Defenders 2014, p. 2) [31].

The vast majority of cases brought against protestors have been unsubstantiated and failed to succeed in court. As an indicative example, one key protest figure has faced nearly 50 proceedings without conviction (Front Line Defenders 2014) [31]. However, even when proceedings have failed to succeed in court, they have the effect of *"deactivat[ing] the uprising and demobilis[ing] the people"* (N_3), in particular by defaming individuals concerned (Front Line Defenders 2014, p. 2) [31].

3.6.3   POLARIZING THE COMMUNITY

Among activists, there is a broad perception that the government, and the company, attempted to polarize members of the community in order to destabilize the protest movement: *"they polarize and they get to the point of the confrontation between the groups that are in favor of the activity and those that are against it"* (P_1). Measures that were apparently implemented to polarize the community have been wide ranging, and according to N_5 even included arming compliant community members: *"The mining company is arming many of the rural workers, for them to confront those who oppose the project."*

Smear campaigns launched against leaders of the protest that aim to defame and delegitimize their efforts have proven the most effective mechanism (Catapa n.d.) [51]. Government rhetoric has consistently characterized protesters opposing major mining projects as "enemies of the state," or people opposed to Peru's development (Sullivan 2014, p. 135) [22]. For instance, former Prime Minister Oscar Valdes described protesters opposing Conga as "anti-investment, anti-development, and opposed to national interest." He proclaimed that "Peruvians need investment to create more jobs. What we don't need is disorder" (Poole and Rénique n.d.) [64].

Similar language has since been reflected in mainstream media reporting in Peru, which has portrayed the opponents to the mine as a minority of violent extremists (Vasquez 2013) [40]. *"Those who oppose the mine are de-legitimized, they are also branded terrorists"* (N_4). This rhetoric is powerful, denoting malicious connotations for an already vulnerable community exercising fundamental rights to oppose a project that would drastically impact their lives and livelihoods. It has had the desired effect of assimilation by the general population: in particular the concept that you are either for, or against development in Peru (Front Line Defenders 2016) [53]. N_3 noted that in Cajamarca, members of the community often perceive protesters as *"someone who is against the wealth of Cajamarca."*

Unsurprisingly, similarly divisive language has been consistently deployed in relation to other mining related conflicts in Peru (Arellano-Yanguas 2012) [1]. Feldman (2014) [16] notes it has become a tool commonly used by government to effect political division.

3.6.4   LACK OF ACCOUNTABILITY

Considering the numerous reported and documented human rights abuses in the context of Minas Conga, the absence of governmental and corporate accountability is remarkable. Contrary to commitments to hold human rights perpetrators accountable for their abuses (Human Rights Council 2013; Voluntary Principles 2000 [45]), no police officer or soldier has been implicated. Incredibly, this lack of accountability has its basis and justification in Peruvian national law. Law No. 30151 of January 2014 provides that members of the national police or armed forces are exempt from criminal responsibility for injury or death caused on duty (Front Line Defenders 2014, p. 3) [31].

3.6.5   DISCUSSION

The analysis of the State and company responses to the protest at Minas Conga demonstrated a tendency for parties to participate, either assertively or complicitly, in human rights abuses against opponents of the mining project. Discussion of the factors that motivated the conduct of the State, the company, and the community can provide insight into why the human rights framework has been ineffective in this specific context.

The State's Failure to Respect and Protect Human Rights

The State's apparent disregard for its human rights commitments in response to the community's mobilization would appear motivated, at least in part, by the importance of mega-mining projects, such as Minas Conga, to Peru's economy. In 2011, the extractive sector generated 11% of Peru's gross domestic product, and accounted for 64% of its exports (Resource Governance, n.d.) [65]. Delays in production have tangible and significant impacts on Peru's economy. An awareness of this sentiment is reflected in the government's rhetoric to convince impacted communities to accept Minas Conga—and when such acceptance was not forthcoming, to justify its subsequent failure to observe its human rights commitments (Sullivan 2014) [22]. N_2 describes this observation, noting *"the big tendency to make the problems of mining activity invisible in contrast to the benefit that it brings to the city or the government."*

   While related to another project, comments from Peru's Interior Minister, Mr. Jose Luis Perez Guadalupe, made after a state of emergency was declared at Las Bambas mine in October 2015, indicate the government considered the economic impact of such projects rendered consultation with the community redundant: *"[the protesters]* ***can't block a project of this dimension,*** *which is an immense investment, the biggest in the past few years"* (Callaghan 2015) [50]. Indeed, the country's commitment to development and economic progress, grounded in resource extraction, has meant that *"the state has opened up, [and is] being as flexible as they can be with the guidelines and norms"* (N_3).

   At the same time, when the Peruvian government's effort to address and placate opponents to large-scale projects have not been effective, it has often faced intense pressure from the mining companies. Multinational firms would urge the State to seek prompt resolution to community opposition, so that operations could proceed or re-commence. On occasion, other actors—such as foreign governments—are introduced to apply further pressure. For instance, a 2009 cable from the United States embassy in Lima outlines implications for US-Peru free trade agreement that were articulated to Peruvian officials, in the event that operations at the Bagua mine did not imminently proceed (Green Left 2014) [54]. Former Peruvian Prime Minister Óscar Valdés alluded to similar or related pressures in comments about Conga, when he referred to the compensation that Peru would have to return if the project did not go ahead (Ponce de Leon 2012) [63].

   In addition to these economic factors and the accompanying external pressures, there are several internal factors that contribute to the government's response observed at Minas Conga.

First, previous authoritarian leaders in Peru demonstrated a tendency to criminalize and suppress opposition, particularly the Fujimori and Garcia administrations in the 1990s and 2000s. Although years have passed since these administrations were in power, their tendencies have left a lasting legacy on government processes (The Economist 2016) [67]—which has been complemented by the public's condemnation of "radical" opposition as a result of Peru's recent exposure to the Shining Path terrorist group. While Peru has actively sought to improve its human rights compliance in recent years, the lasting legacy continues to influence contemporary government decision-making. In particular, by contributing to its bias against the interests of vulnerable populations.

There is also a notable disconnect between Peru's central government based in Lima and the many rural and indigenous communities situated across the country. This disconnect is both geographical and cultural, and manifests in the divergent values of the State and the individual communities (Sullivan 2014) [22]. Generally, these rural and indigenous communities do not possess meaningful representation or political power to influence central government decisions, making it easy for government administrations to ignore their interests. Further, decision-makers within the central government recognize that the State lacks the financial and human resources, as well as the infrastructure required to facilitate discussions with rural and indigenous communities about their concerns and grievances. The perception that resolving these concerns cannot easily be achieved through consultation leads some decision-makers to the conclusion that heavy-handed and repressive responses are necessary. For example, reports suggested that the government declared a state of emergency on December 5, 2011 in order to prevent mobilizations from spreading *across the country* (Poole and Rénique n.d.) [64]. Peruvian activist Jose de Echave noted that such declarations are the "*clearest sign the government does not know how to deal with conflict*" (The Guardian 2015) [69].

As a net result of its openly repressive tactics, Peru's government has demonstrated its willingness to subjugate the rights of some of its most vulnerable and powerless communities in favor of the development of mineral resources in the country (Sullivan 2014, p. 135) [22]. This means that peasant communities impacted by mining projects like Conga are left unprotected against the combined forces of the State and mining corporations. According to C_4, "*as a citizen …we don't matter to the state,*" while C_5 elaborated further: "*the state …govern for the transnationals and not in favor of our country.*"

3.6.6 THE INADEQUACY OF BUSINESS COMMITMENTS TO RESPECT HUMAN RIGHTS

Newmont equally demonstrated indifference to its public human rights commitments through its conduct during the Minas Conga project. On its face, Newmont's conduct and interactions with the community appear primarily motivated by its desire for immediate results. This is evidenced by the company's active efforts to circumvent the community consultation during the EIA process, its use of intimidation during the land acquisition process, and its apparent (whether tacit or otherwise) support for heavy-handed government responses to large-scale mobilizations at the site.

When a State's regulation of mining is not particularly strong, and their economy is reliant upon mining, business human rights instruments contemplate that the company should assume a greater human rights burden (Slack 2012) [21]. In the case of Minas Conga, under its business commitments to respect human rights, Newmont might have been expected to take measures to prevent or mitigate the human rights abuses committed by the State forces (Guiding Principles 2011, principle 13) [24], including using its influence to restrict the use of force to avoid human rights violations (Voluntary Principles 2000) [45].

Unfortunately, as observed at Conga, companies often choose not to intervene, even when it possesses knowledge that State forces are violating human rights and repressing protest. Instead, Newmont remains resolute that its decisions and behavior reflected its neutrality.

"We do not get in the middle of the government's decision making around how they choose to conduct themselves when they make decisions to have an intervention" (M).

Newmont's inaction, while it had knowledge that Peruvian security forces were likely commiting human rights violations, is inconsistent with its human rights commitments, which call on companies to ensure the rights of vulnerable individuals and groups affected by their activities (Guiding Principles 2011, principle 19) [24].

Newmont's unwillingness to intervene highlights the aspirational nature of the human rights framework, which does not adequately account for the pragmatic reality that corporate behavior is generally driven by commercial interests. Even despite the increasing influence of corporate social responsibility on extractive sector corporate rhetoric, "social" policies generally remain motivated by quantifiable costs to the project in their absence (Harvey 2014, p. 7) [13].

The failure of the business commitments to human rights to protect vulnerable communities, such as those impacted by Minas Conga, can be attributed to several factors (GRUFIDES 2013) [32]. First, the relevant documents are drafted in broad and vague terms, meaning it is often unclear what steps or measures a company should take to meet their commitments. Second, the documents are self-regulated and not legally binding (RESOLVE 2016, p. 13) [39], which inextricably links their effectiveness to the willingness of the company to enact them. Finally, the consequences for failing to comply with the commitments are virtually non-existent. This reality was demonstrated by the Voluntary Principles Initiative Secretariat's failure to respond to a complaint lodged by LAMPP, which called for Newmont's expulsion from the group, citing the company's role in human rights abuses at Minas Conga (LAMMP 2014, p. 6) [37].

Yet, Newmont retains its resolute public commitment to respecting human rights. The company consistently reaffirms its commitment to the key human rights standards in public facing documents and continues to participate in the Voluntary Principles group at the international level (RESOLVE 2016, p. 14 [39]; Newmont 2014a [61]). Minera Yanacocha also coordinates "Grupo Impulsor," a working group that promotes the implementation of the Voluntary Principles in Peru (RESOLVE 2016, p. 14) [39]. This group works with the Peruvian government and regularly provides information to police about the Voluntary Principles (RESOLVE 2016, p. 36) [39]. According to *M*, Newmont *"demands the opportunity to provide awareness on human rights, and what our standards and our obligations are, and do our level of diligence to ensure that all of those conditions are being met."*

It is unclear whether Newmont's recent efforts constitute a change in attitude toward their human rights commitments in response to the negative global reception of its involvement in the Minas Conga, or whether these commitments are merely superficial, in order to legitimize claims of human rights compliance.

3.6.7   A DIFFERENT CONCEPT OF "DEVELOPMENT"

When considered carefully, there is a profound incompatibility between the motives of the State (and for that matter, the company as well) and the protesters. On one hand, the government and the company promote the primacy of development, stemming largely from economic incentives (Dagnino 2005) [8]. From the broader community perspective, development should be pursued within a framework of strong environmental protection and the maintenance of traditional livelihoods. As a general principle, communities will reject attempts to interfere with their existing livelihoods and the environment, particularly fresh water sources (De Echave 2008 [9]; Bebbington et al. 2008 [2]).

Despite the mining-related conflict across the country, protesters have expressed that they are not strictly anti-mining or anti-development. Instead, they protest for "responsible mining practices" on two fronts: environmental responsibility and social responsibility (Front Line Defenders 2016) [53]. N_2 described that the community *"had the view that a company could eventually develop activities without violating the rights, or affecting the environment."*

Fears about water sources drove community members' negative reception of the project: *"[water] means death [or life] for them …if their water supply is contaminated and dried up then how could they survive"* (P_2). The significance of the water to the local community was exemplified during the National March in Defense of Water in February 2012. Protesters marched from Conga to Lima over eight days to present Congress a legislative proposal for an outright ban on mining in headwaters (Zibechi 2012) [74].

A lack of insight and understanding from company and State into the deep-seated community traditions and the significance of local water sources contributed to failed community engagement attempts earlier in the mine's life. For instance, Newmont attempted to elicit the community's support through gifts and promises of jobs: *"they arrived to the school with uniforms, with backpacks, educational materials to the teachers, with gifts"* (P_1). In subsequent attempts to appease the local community, the government announced a $1.6 billion program of social and infrastructure investments in Cajamarca in January 2012 (Moffett and Dube 2012) [70].

These measures were ultimately rejected in favor of saving the lakes, the environment, and traditional livelihoods. This context frames the current impasse at Conga, founded in the division between the diverging values: one that values economic development, and another that values nature and their traditional existence. If there is not significant compromise from either or both parties, these opposing positions are irreconcilable, which was perhaps reflected by the recent decision to indefinitely suspend the Conga project (Newmont 2016a) [58].

3.6.8 THE IMPORTANCE OF ACTIVISM AND RAISING AWARENESS

The discussion above has demonstrated that in the current geopolitical context, neither State nor business commitments to respect and uphold human rights will reliably ensure the protection of community members impacted by mining projects in Peru (and by extension, other developing countries across the world that are economically reliant upon the extractive sector). However, the growing international awareness and criticism from NGOs and human rights bodies have contributed to a shift in approach toward protest at Conga.

For instance, in 2013 Newmont commissioned a "Listening Study" in an attempt to understand concerns of the community (CSRM 2013) [36], and an "Independent Fact Finding Mission" this year into Chaupe family dispute (RESOLVE 2016) [39]. Equally, the Peruvian government has established the Office for National Dialogue and Sustainability as a means to democratically resolve conflicts (OECD 2016) [38]. This willingness to "listen" demonstrates positive steps toward human rights accountability on the part of the government and the operator.

3.7 CONCLUSION

A study of the conflict at Conga has revealed that both international and business commitments to human rights fail (in practice) to adequately protect the rights of communities impacted by mining projects in Peru.

The conflicting priorities of the State and company, which favored economic development and production on the one hand, and the impacted community, which favored protection of the environment and retention of their traditional livelihoods, resulted in an impasse that ultimately resulted in resolute community opposition to the mine. Unfortunately, rather than engage in constructive dialogue with opponents to the mine, the State and the company chose to ignore concerns and repress the subsequent community mobilizations.

The apparent disregard for human rights commitments can be explained in part by the significant macro-economic benefits of large mining investments such. At Minas Conga, this "business first" approach to a very complex situation inflamed relations between the respective parties, imposing significant costs upon the community (through the suppression of their rights). As a result, the project was suspended as of April of 2016.

Despite the staunch opposition to Conga, the study revealed that communities are willing to engage in discussions about resource extraction projects, provided that these communities are

- accepted as legitimate stakeholders with the right to veto projects that fail to guarantee socially and environmentally responsible practices; and

- engaged in every stage of the life-cycle of the mine.

Therefore, in the interest of developing mineral resources adjacent to rural and indigenous communities in Peru and Latin America, the mining industry and state governments should work together to develop processes and strategies that ensure environmentally safe mining practices, as well as zero tolerance of human rights abuses—including the use of violence and repression to silence opposition.

3.8   FUTURE WORK

This study has highlighted how, at times, economic interests can undermine the corporate and State commitments to respect and uphold human rights standards (in this case, for protesters opposed to a large-scale mine projects in Peru). The growing body of awareness, criticism, and rejection of these practices—from NGOs, human rights bodies, and community groups—has attracted global attention and delivered incremental steps toward human rights accountability. Further contributions to the awareness of these practices are essential to the continuing growth of human rights compliance within the mining industry.

For mining companies, the development of more considerate and nuanced community engagement strategies must be a priority, if they wish to have social approval of their projects that will ultimately lead to a more stable and sustainable future for mining projects and the communities they impact.

3.9   PEER-REVIEWED ARTICLES

[1] Arellano-Yanguas, J. (2012). Mining and conflict in Peru: Sowing the minerals, reaping a hail of stones. In A. Bebbington, (Ed.), *Social Conflict, Economic Development and Extractive Industry*, pages 89–111, Routledge, New York. 55, 70

[2] Bebbington, A., Bebbington, D. H., Bury, J., Lingan, J., Munoz, J. P., and Scurrah, M. (2008). Mining and social movements: Struggles over livelihood and rural territorial development in the Andes. *World Development*, 36(12):2888–2905. DOI: 10.1016/j.worlddev.2007.11.016. 54, 56, 74

[3] Bebbington, A. (2012). Extractive industries, socio-economic conflicts and political economic transformations in Andean America. In A. Bebbington, (Ed.), *Social Conflict, Economic Development and Extractive Industry*, pages 3–26, Routledge, New York. 56

[4] Bebbington, A. and Williams, M. (2009). Water and mining conflicts in Peru. *Mountain Research and Development*, 28(3):190–195. DOI: 10.1659/mrd.1039. 56

[5] Braaten, R. H. (2014). Land rights and community cooperation: Public goods experiments from Peru. *World Development*, 61:127–141. DOI: 10.1016/j.worlddev.2014.04.002. 56

[6] Bury, J. and Kolff, A. (2002). Livelihoods, mining and peasant protests in the Peruvian Andes. *Journal of Latin American Geography*, 1(1):3–16. DOI: 10.1353/lag.2007.0018. 56

[7] Crabtree, J. and Crabtree-Condor, I. (2012). The politics of extractive industries in the Central Andes. A. Bebbington, (Ed.), *Social Conflict, Economic Development and Extractive Industry*, pages 67–88, Routledge, New York. 55, 56

[8] Dagnino, E. (2005). Meanings of citizenship in Latin America. *IDS Working Paper*. 258 Brighton, Institute for Development. DOI: 10.1080/08263663.2006.10816901. 74

[9] De Echave, J. (2008). *Diez Años de Mineria en el Perú*. Lima, Cooperaccion. 74

[10] De la Flor, P. (2014). Mining and economic development in Peru: A time of resurgence. *Revista – Harvard Review of Latin America*. https://revista.drclas.harvard.edu/book/mining-and-economic-development-peru 53, 55, 58

[11] Dolan, C. and Rajak, D. (2016). *The Anthropology of Corporate Social Responsibility*. Berghahn, New York. DOI: 10.2307/j.ctvgs09h2.5. 65

[12] Esteves, A. (2008). Mining and social development: Refocussing community investment using multi-criteria decision analysis. *Resources Policy*, 33(1):39–47. DOI: 10.1016/j.resourpol.2008.01.002. 54

[13] Harvey, B. (2014). Social development will not deliver social licence to operate for the extractive sector. *The Extractive Industries and Society*, 1:7–11. DOI: 10.1016/j.exis.2013.11.001. 53, 73

[14] Hilson, G. (2012). Corporate social responsability in the extractive industries: Experiences from developing countries. *Resources Policy*, 37:131–137. DOI: 10.1016/j.resourpol.2012.01.002. 53

[15] Holden, W. and Jacobson, R. D. (2007). Mining amid armed conflict: Nonferrous metals mining in the Philippines. *The Canadian Geographer*, 51(4). DOI: 10.1111/j.1541-0064.2007.00193.x. 56

[16] Feldman, I. A. (2014). *Rethinking Community from Peru: The Political Philosphy of José Maria Arguedas*. University of Pittsburgh Press, Pittsburgh. DOI: 10.2307/j.ctt9qh5w0. 70

[17] Li, F. (2009). *Unearthing Conflict*. Duke University Press, Durham. DOI: 10.1215/9780822375869. 64

[18] Martin, B. (1986). Protest in a liberal democracy. *Seminar on the Right of Peaceful Protest*, pages 94–117. 62

[19] O'Flaherty, M. (2014). Effective measures and best practices to ensure the promotion and protection of human rights in the context of peaceful protests: A background paper. *Irish Centre for Human Rights*. 62

[20] Ruggie, J. G. (2007). Business and human rights: The evolving international agenda. *KSG Working Paper*, no. RPW07–029. DOI: 10.2139/ssrn.976547. 63

[21] Slack, K. (2012). Mission impossible?: Adopting a CSR-based business model for extractive industries in developing countries. *Resources Policy*, 37:179–184. DOI: 10.1016/j.resourpol.2011.02.003. 73

[22] Sullivan, L. (2014). Getting to the bottom of extractive capitalism. *Policy and Practice: A Development Education Review*, pages 124–143. 68, 70, 71, 72

[23] Triscritti, F. (2013). Mining, development and corporate-community conflicts in Peru. *Community Development Journal*, 48(3):437–450. DOI: 10.1093/cdj/bst024. 56, 57, 58, 59, 60, 61, 64, 68

3.10   NON-GOVERNMENTAL ORGANIZATION REPORTS

[24] Armstrong, R., Baillie, C., Fourie, A., and Rondon, G. (2014). Mining and community engagement in Peru: Communities telling their stories to inform future practice. *International Mining for Development Centre*, University of Western Australia (IM4DC). 55, 56, 59, 60, 61, 63, 64, 66, 68, 73

[25] Columbia Law School Human Rights Clinic (2015). Conga no va: An assessment of the Conga mining project in light of World Bank standards. *Columbia Law School Human Rights Clinic (CLSHRC)*, Columbia University. web.law.columbia.edu/sites/default/files/microsites/human-rights-insitute/conganova_english.pdf 54, 57, 59, 60, 61, 65, 68

[26] Defensoria del Pueblo (2010). Reporte mensual no. 80 conflictos sociales. Defensoria del Pueblo, Lima. 57

[27] Defensoria del Pueblo (2012). Violencia en los conflictos socialies. Informe defesorial no. 156, Defensoria del Pueblo, Lima. 54, 57

[28] Delgado, A. and Baracco, D. (2019). Mining in Peru: Overview. *Practical Law*. Thomson Reuters. https://uk.practicallaw.thomsonreuters.com/w-008-1009?transitionType=Default&contextData=(sc.Default)&firstPage=true&comp=pluk&bhcp=1 65

[29] Earth Rights International (2014a). Factsheet: Campo-Alvarez vs. Newmont Mining Corp. https://www.earthrights.org/sites/dedefault/files/documents/Factsheet-Campos-Alvarez-v-Newmont.pdf 59, 60, 61, 66, 68

[30] Earth Rights International (2014b). U.S. federal court action requests information from Newmont regarding repression of protests at its Conga mine project. https://www.earthrights.org/media/us-federal-court-action-requests-information-newmont-regarding-repression-protests-its-conga 55

[31] Front Line Defenders (2014). Environmental rights defenders at risk in Peru. https://www.frontlinedefenders.org/en/statement-report/environmental-rights-defenders-risk-peru 68, 69, 70

[32] GRUFIDES (2013). Police in the pay of mining companies: The responsibility of Switzerland and Peru for human rights violations in mining disputes. assets.gfbv.ch/downloads/resport_english_def_2_12_13.pdf 66, 67, 73

[33] Human Rights Watch (2016). Peru: Events of 2015. *Human Rights Watch*. https://www.hrw.org/world-report/2016/country-chapters/peru 57, 67

[34] Human Rights Watch (2015). *World Report 2014*. https://www.hrw.org/sites/default/files/reports/wr2015.pdf [2 May 2016]. 55

[35] International Union for Conservation of Nature (IUCN) (2016). Communal lands and protected areas in Peru. https://www.iucn.org/sites/dev/files/content/documents/tger_peru_final-english.pdf 65

[36] Kemp, D., Owen, J., and Arbelaez-Ruiz, D. (2013). Listening to the city of Cajamarca. *Centre for Social Responsibility in Mining (CSRM)*, University of Queensland. https://www.csrm.uq.edu.au/publications/listening-to-the-city-of-cajamarca-a-study-commissioned-by-minera-yanacocha-final-report 54, 55, 56, 57, 59, 60, 61, 75

[37] Latin American Mine Monitoring Programme (LAMMP) (2014). Maxima acuna chaupe. lamp.org/wp-content/uploads/2014/09/vvpma1.pdf 66, 73

[38] Organisation for Economic Co-operations and Development (2016). Environmental performance review: Peru, highlights and recommendations. *Organisation for Economic Co-operations and Development (OECD)*. https://www.oecd.org/environment/country-reviews/16-00312-environmental%20performance%20review-peru-web.pdf 75

[39] RESOLVE (2016). Tragadero Grande: Land, human rights, and international standards in the conflict between the Chaupe family and Minera Yanacocha. www.resolv.org/site/yiffm/files/2015/08/YIFFM-report_280916-Final.pdf 59, 60, 61, 65, 66, 73, 74, 75

[40] Vasquez Chuquilin, M. (2013). La criminalizacion de la protesta social en el Perú: Un análisis a la luz del caso Conga en Cajamarca. *GRUFIDES*, Cajamarca. 70

[41] World Bank (2005). Wealth and sustainability: The environmental and social dimensions of the mining sector in Peru. Washington, DC, World Bank. http://documents.worldbank.org/curated/en/2005/12/7041000/wealth-sustainability-environmental-social-dimensions-mining-sector-peru-vol-2-2-main-report 55

3.11   TREATIES, UN DOCUMENTS, DOCUMENTS FROM INTERNATIONAL ORGANIZATIONS

[42] Congreso de la República del Perú (1993). *Political Constitution of Peru.*

[43] Inter-American Commission on Human Rights (2014). Resolución 9/2014: Lideres y lideresas de comunidades campesinas y rondas campesinas de Cajamarca respecto de la Republica de Perú. http://www.oas.org/es/cidh/decisiones/pdf/2014/MC452-11-ES.pdf 66

[44] *International Covenant on Civil and Political Rights* (1966). United Nations, Treaty Series, 999:171.

[45] *The Voluntary Principles on Security and Human Rights* (2000). Foley Hoag LLP, the secretariat for the voluntary principles on security and human rights. http://www.voluntaryprinciples.org/wp-content/uploads/2013/03/voluntary_principles_english.pdf 63, 66, 70, 73

[46] United Nations (2000). *The Ten Principles of the UN Global Compact*, United Nations Global Compact. https://www.unglobalcompact.org/what-is-gc/mission/principles 63

3.12   NEWS ARTICLES AND OTHER PUBLICATIONS

[47] AAP (2013). Peruvians protest Newmont mine plans. *9Finance*. http://finance.nine.com.au/2016/10/21/20/39/peruvians-protest-newmont-mine-plans 65, 68

[48] Al Jazeera (2012). Peru protests ensue despite emergency decree. *Al Jazeera*. http://www.aljazeera.com/news/americas/2012/07/20127513330781731.html 69

[49] BBC (2015). Peru anti-mining protest sees deadly clashes. *BBC News*. http://www.bbc.com/news/world-latin-america-34389803 69

[50] Callaghan, H. (2015). Peru declares state of emergency. *Activist Post*. http://www.activistpost.com/2015/10/peru-declares-state-of-emergency.html 71

[51] Catapa (n.d.) The Conga mega project. *Catapa*. catapa.be/en/cases/peru/conga/conga-mega/project 69, 70

[52] El Comercio (2012). Protestas mineras en Cajamarca sumaron quinto fallecido en dos días 2012, *El Comercio*. http://elcomercio.pe/sociedad/lima/protestas-antimineras-cajamarca-sumaron-quinto-fallecido-dos-dias-noticia-1437526?ref=flujo_tags_327504&ft=nota_50&e=titulo 68

[53] Front Line Defenders (2016). Peru—mining and human rights defenders. https://www.frontlinedefenders.org/en/statement-report/environmental-rights-defenders-risk-peru 67, 70, 74

[54] Green Left (2014). Peru: Wikileaks cables shed light on U.S. massacre role. *Green Left Weekly*. https://www.greenleft.org.au/content/peru-wikileaks-cables-shed-light-us-massacre-role 71

[55] International Council on Mining and Minerals (2015). Good practice. *International Council on Mining and Metals (ICMM)*, 13(1). https://www.icmm.com/document/8633 63

[56] Internacional de Resistentes a la Guerra (2015). Mining, militarization and criminalization of social protests in Latin America. *Internatcional de Resistentes a la Guerra*. http://www.wri-irg.org/es/node/24540 56

[57] Latin Resources (n.d.) Overview of process for grant of a mining concession in Peru. *Latin Resources Peru*. http://latinresources.com.au/overview_of_process_for_grant_of_a_mining_concession_in_peru 65

[58] Newmont (2016a). Form 10-K, annual report pursuant to section 13 or 15D of the securities exchange act of 1934. *United States Securities and Exchange Commission*. https://www.sec.gov/Archives/edgar/data/1164727/000155837016003258/nem-20151231x10k.htm 59, 60, 61, 75

[59] Newmont (2016b). Chaupe land case information update: April 27, 2016. *Newmont Mining*. http://s1.q4cdn.com/259923520/files/doc_downloads/Chaupe/Chaupe-Issue-Stakeholder-Update-27-April-2016.pdf 66

[60] Newmont (2015). Annual report to the voluntary principles on security and human rights. *Newmont Mining*. http://www.voluntaryprinciples.org/wp-content/uploads/2015/10/Annual-Report-Voluntary-Principles-Newmont-Mining-Corporation_August-2015.pdf 59, 60, 61, 64

[61] Newmont (2014a). Sustainability and stakeholder engagement policy. *Newmont Mining*. http://s1.q4cdn.com/259923520/files/doc_downloads/newmont_policies/Policy_Sustainability-StakeholderEngagement_28Apr2014.pdf 59, 60, 61, 64, 73

[62] Peru Support Group (n.d.) Criminalisation of social protest. *Peru Support Group*. http://www.perusupportgroup.org.uk/peru-human-rights-social-protest.html

[63] Ponce de Leon, R. (2012). Failure to develop Minas Conga could leave huge compensation claim—PM. *Business News America*. http://www.bnamericas.com/news/mining/failure-to-develop-minas-conga-could-leave-huge-compensation-claim-pm 71

[64] Poole, D. and Rénique G. (n.d.) Peru: Humala takes off his gloves. *North American Congress for Latin America*. https://nacla.org/article/peru-humala-takes-his-gloves DOI: 10.1080/10714839.2012.11722104. 57, 69, 70, 72

[65] Resource Governance (n.d.) Peru. *Natural Resource Governance Institute*. http://www.resourcegovernance.org/our-work/country/peru 71

[66] Reuters (2011). Peru urges new water plan for $4.8 bln Conga mine. *Reuters*. http://in.reuters.com/article/peru-newmont-idINN1E7A30SX20111104 57

[67] The Economist (2016). From conflict to co-operation; Mining in Latin America. *The Economist*, p. 41. www.economist.com/news/americas/21690100-big-miners-have-better-record-their-critics-claim-it-up-governments-balance 54, 56, 72

[68] The Guardian (2011). Peru declares state of emergency to end protests over mine. *The Guardian*. https://www.theguardian.com/world/2011/dec/05/peru-state-of-emergency-protests-mine 69

[69] The Guardian (2015). What is Peru's biggest environmental conflict right now?. *The Guardian*. https://www.theguardian.com/environment/andes-to-the-amazon/2015/jun/08/tia-maria-perus-biggest-environmental-conflict-right-now 72

[70] Moffett, M. and Dube, R. (2012). Peru tries to appease mine protesters. *Wall Street Journal*. http://www.wsj.com/articles/SB10001424052970204409004577159030316618706 74

[71] Mora, R. (2015). Foreign mining companies hire Peru's police as private security. *Telesur*. http://www.telesurtv.net/english/news/Perus-Police-Criticized-for-Private-Financing-from-Business-20150507-0028.html 66, 67

[72] Triscritti, F. (2012). The criminalization of anti-mining social protest in Peru. *Columbia University Earth Institute*. http://blogs.ei.columbia.edu/2012/09/10/peru-mining/ 56

[73] United States Geological Survey (2015). Commodity statistics and information. minerals.er.usgs.gov/minerals/pubs/commodity/ 55

[74] Zibechi, R. (2012). Latin America: A new cycle of social struggles. *North American Congress for Latin America*. https://nacla.org/article/latin-america-new-cycle-social-struggles DOI: 10.1080/10714839.2012.11722089. 74

[75] ICMM International Council of Mining and Minerals (2013). *Community Development Toolkit*. http://www.icmm.com/community-development-toolkit 54

CHAPTER 4

Exploring the Notion of Socially Just Mining Through the Experiences of Five Indigenous Women from Latin America

Kylie Macpherson

4.1 INTRODUCTION

Across the world, metals have been mined for millennia. However, in past decades with the arrival of new technology the nature of mineral extraction has undergone significant transformation, shifting toward large, mechanized, open-pit operations that command unprecedented transnational capital flows (Armstrong, Baillie, Cumming-Potvin 2014) [5]. The availability of cheap fuel together with the use of new technology made it possible to develop new strategies and techniques for recovering low-grade ore, previously considered uneconomical to extract (BNamericas 2011) [15]. Latin America's wealth of natural resources, cheap labor, and a welcoming political environment are factors that contributed to an unprecedented mining growth bringing great wealth and prosperity for some countries and corporations but also great disruption of livelihood and suffering for others (Bebbington 2012 [12]; LAMMP 2016[53]).

Support of the modern mining paradigm is centered on arguments of "mining for development" and indicators such as GDP and economic growth, which tell a story of community benefits and poverty alleviation (Broad 2014 [18]; Graulau 2008 [44]). However, the nature and distribution of these benefits has been questioned, and communities remain apprehensive about the contribution of mining projects to local development (Vittor 2014) [84]. Social conflict, loss of collective rights, eviction of vulnerable groups from their land, and environmental degradation have tainted the mining legacy (Dajer 2015 [27]; Pannwitz 2014 [90]). Heightened awareness of these issues has exposed miners to a new consumer-critique, driving the business case for "Corporate Social Responsibility" (CSR) and "social license to operate" (Harvey 2013 [48]; Kemp and Owen 2013 [50]). CSR is taken here to mean that: *"corporations have a degree of responsibility not only for the economic consequences of their activities, but also for the social and environmental implications"* (Australian Human Rights Commission 2008 [9]).

Despite the proliferation of standards for best practice as well as UN rights' frameworks, local communities continue to see human rights abuse and severe damage to natural habitats as an inevitable consequence of irresponsible mining projects (Council on Hemispheric Affairs 2014) [24]. Perception of mining as a threat to people and the environment made Costa Rica's congress impose a ban on all gold mining in 2010 (The Costa Rica News 2018) [75]. In March 2020, lawmakers in El Salvador overwhelmingly passed legislation imposing a "blanket ban" on all metals mining in the small, densely populated Latin American nation (Palumbo and Malkin, 2017) [63]. The ruling is the product of over 11 years of grassroots community advocacy including education, street blockades, and rallies in defense of indigenous territories. This outcome sends a clear message to mining organizations that the prevailing "audit culture" and associated withols for ensuring positive and equitable social benefits with minimal environmental impact do not always lead to responsible mining. Audit culture refers to the industry standard practice of assessing outcomes against a pre-determined benchmark value in order to measure and justify corporate actions (Kemp, Owen and van de Graaf 2011) [50].

4.1.1   THE LATIN AMERICAN CONTEXT

The shift toward large-scale, capital-intensive mining and foreign direct investment (FD) is well documented throughout Latin America. Led by the "Chilean Miracle,[1]" nations across the region have undergone a common restructuring of economic and political policy since the early 1990s. The policy shifts have lifted domestic protectionist laws in order to encourage and protect FDI (Bebbington 2012 [12]; Bury 2005 [19]). Key restructuring activities (Table 4.2) have reduced the presence of the state in economic and social affairs and established self-regulating supply, demand, and pricing mechanisms (Bury 2005) [19].

The result has been a rapid integration into the global economy where mining is perceived as central for growth and export-led earnings (Bury 2005 [19]; The Economist 2016 [74]). The opening of state-protected economies such as Guatemala, Ecuador, Honduras, and Peru has resulted in a common increase in GDP per capita from 1990–2005. Further FDI has increased to levels consistent with major open global economies—typically running at 2–5%—and exports comprise a larger portion of overall GDP (Table 4.2) (World Bank Group 2017) [88].

However, it is evident that the reception of neoliberal reform has not been equal for all. The Gini Coefficient, an index measure of income inequality within a nation, has increased across all countries (except Guatemala) and all four remain significantly higher than that of other

---

[1]The "Chilean Miracle" refers to the reorientation of the Chilean economy during Augusto Pinochet's dictatorship 1973–1990. Pinochet led Chile through a free-market reform that resulted in sustained and high rates of economic growth, a process dubbed "a success story, a model for other countries to follow" (Richards 2013, p. 1) [69].

Table 4.1: Acronyms

Acronym	Description
AusIMM	Australian Institute of Mining and Metallurgy
CEME	Civil, Environmental, and Mining Engineering
COPINH	Consejo Cívico de Organizaciones Populares e Indígenas de Honduras— (National Council of Popular and Indigenous Organizations of Honduras)
CSR	Corporate Social Responsibility
DIDO	Drive-in-Drive-out
EIA	Environmental Impact Assessment
FDI	Foreign Direct Investment
FIFO	Fly-in-Fly-out
GDP	Gross Domestic Product
HRD	Human Rights Defender
LAMMP	Latin American Mining Monitoring Programme
LPSDP	Leading Practice Sustainable Development Program
MIGA	Multilateral Investment Guarantee Agency
NGO	Non-Governmental Organization
OPIC	Overseas Private Investment Corporation
ULAM	Unión Latinoamericana de Mujeres (Latin American Women's Union)
UN	United Nations
UNGP	United Nations Guiding Principles
UNDRIP	United Nations Declaration on the Rights of Indigenous Peoples
UWA	University of Western Australia

globalized economies—indicating that wealth and growth has been concentrated on certain groups or individuals (World Bank Group 2017) [88].

Today, Latin American nations represent a diverse political construct in various stages of neoliberal and post-neoliberal reform where citizens have began to "claim back their rights" to effectively harness the benefits (and costs) of mining (Bebbington 2012) [12]. Disputes over extraction of natural resources as the best model to tackle poverty and achieve better societies (Denniss 2014) [28] minimize the inequitable distribution of benefits and externalities in the context of mining are common and correlated with the sector's contribution to GDP (%).

Table 4.2: Policies of Latin American neoliberal reform

Latin American Neoliberal Reform 1990-2005
1. Opening of all sectors of the economy to FDI and lifting of restrictions on remittance of profits, dividends, royalties, access to domestic credit, and acquisition of supplies and technology abroad;
2. Offering of tax stability packages for foreign investors for terms of 10–15 years, privatization programs offering international investment opportunities, and improved competition;
3. Ratifying of bilateral and multilateral investment-guarantee treaties, such as Multilateral Investment Guarantee Agency (MIGA) convention and the Overseas Private Investment Corporation (OPIC) accords;
4. Revision of land-tenure rights, e.g., 1996 National Mining Cadastre Law in Peru Law 26615.

Table 4.3: Economic indicators of Latin America 1990–2005 (Source: data.worldbank.org, 2017)

	Guatemala		Ecuador		Honduras		Peru	
	1990	2005	1990	2005	1990	2005	1990	2005
GDP per capita (2010 US$)	2,171	2,620	3,721	4,287	1,549	1,915	2,679	3,830
Gini Coefficient	59.6	54.9	51.7	54.1	57.4	59.5	44.7	51.8
FDI (% of GDP)	0.62	1.98	0.83	2.29	1.43	6.21	0.16	3.39
Exports (% of GDP)	18.8	25.1	22.8	27.6	37.2	59.0	13.0	26.9

4.2   PROBLEM IDENTIFICATION

This chapter explores perceptions of socially just mining through the experiences of Latin American indigenous women human rights defenders (WHRDs) directly impacted by mining developments. Specifically, it examines first the ways in which CSR safeguards are inappropriate for the local context of indigenous communities and how they fail to uphold basic human rights for vulnerable groups; and second it looks into the consequences for communities of traditional models of CSR that do not necessarily use a human rights approach. This is why the UN Guiding Principle (UNGP) framework which distinguishes between the state's duty to protect and the corporations' responsibility to respect is such an essential guide for corporations. Future implementation of the Principles is likely to have significant impact on consultations and environmental impact assessment (EIA), calling for greater understanding of the misalignment between corporations and the communities in which they operate.

Theories of social justice are well documented in the literature—the discussion herein will consider historical theories of social justice (Mill 1863 [58]; Rawls 1973 [67]) and explore recent ideas that apply these theories in a modern, globalized context (Franklin 1999 [36]; Lettinga and van Troost 2015 [55]).

John Stuart Mill's (1863) theory of Utilitarianism accepts utility as the foundation of justice as an ethical judgment. In his 1863 publication, *Utilitarianism*, Mill states:

> "…*actions are right in proportion as they tend to promote [utility], wrong in proportion as they tend to produce the reverse*," where utility refers to "*pleasure and the absence of pain*" (Mill 1863, p. 5) [58].

Further, in applying utilitarianism, Mill identifies the need for moral human conduct in assessing one's utility, asking:

> "*What means do we have for deciding which is the more acute of two pains, or the more intense of two [pleasures], other than the collective opinion of those who are familiar with both?*" (Mill 1863, p. 8) [58].

By considering who may derive the greater utility [benefits] from an action, Mill argues that the perfectly "just" conception of utility (and morality) is the condition whereby "the greatest amount of happiness altogether … as far as possible from pain" exists.

John Rawls' *A Theory of Justice* (1973) [67] highlights issues in applying Mill's maximum utility for social justice theory, specifically its lack of consideration for "*how this sum of satisfactions is distributed among individuals*" (Rawls 1973, p. 26) [67]. Identifying that "*the violation of the liberty of a few might not be made right by the greater good shared by many*," Rawls developed his theory of "Justice as Fairness" deriving the Liberty Principle whereby:

> "*Each person is to have an equal right to an adequate scheme of equal basic rights and liberties compatible with a similar scheme of liberties for all*" (Rawls 1999, p. 52–53) [68].

This principle, Rawls argues, is essential in the construction of the political concept of justice. In exercising Justice as Fairness, Rawls outlines an abstract mental device—the "Original Position" that asks us to consider:

> "*If all parties entering a contract were to cast a 'veil of ignorance' over their original position, would these parties still enter into the contract aware of the outcomes of 'life and society' imposed to all parties*" (Rawls 1991, 24–25) [68].

Adopting a veil of ignorance over our "original position," e.g., our economic endowments, gender, race or religion, Rawls argues, will allow impartial reasoning about the fundamental principles of justice when entering a social contract.

A problem with *Justice as Fairness* is that in was conceived in the context of a closed economy where the "who" of a given social contract is easily identified—our neighbors, the society downstream, our children, and so forth—and focus may be afforded to the "what" of social outcomes. Ursula Franklin's *A Real World of Technology* (1990) [36] highlights issues pertaining to the application of Rawls' justice principled in a globalized world where benefit (and cost) flows transcend borders into new, unfamiliar frontiers. Here, not only the "what" but also the "who" of social impacts can be confounding. Franklin predicts a renewed attention to justice in the receiving of technology in a globalized world and proposes greater attention to the language of the discourse, stating:

> *"Whenever someone talks to you about the benefits and costs of a particular project, don't ask 'What benefits?' ask 'Whose benefits?' and 'Whose costs?'"* (Franklin 1990, p. 124) [36].

In this way, a new accounting system that focuses on the distribution of benefits, enriched by the *"rephrasing of an observation in line with a perspective from the receiving end of technology"* (Franklin 1990, p. 124) [36], is proposed. Further, Franklin highlights, globalized *"production models"* are *"perceived and constructed without links to a larger context"* allowing the leveraging or application of *"a particular model in a variety of situations."* As a result, externalities of a production model are often considered irrelevant to the activity itself, becoming the business of someone else. This separation of owner from impact, in the context of mining, has resulted in failure to consider external and interactive effects of a project, leading to inadequate modeling of CSR.

These varying ideas are inherent to social justice as a *normative ethical judgment*, rather than an empirically verifiable outcome or 'ideal' construct of society (Cramme and Diamond 2009) [26]. For the purpose of the research herein, social justice will draw on Rawls' theory of justice as *"an equal right to an adequate scheme of equal basic rights and liberties compatible with a similar scheme of liberties for all"* (Rawls 1973, p. 52) [67].

4.2.1   SOCIAL JUSTICE IN THE PRAXIS OF MINING

Similarly, social justice *praxis*[2] in the context of mining can also mean different things to different people, even from relatively similar backgrounds. This may be attributed, in part, to the cause of the author: some argue for a defined set of industry guidelines (Harvey 2013 [48]; LPSDP 2016 [8]; Newmont Mining Corporation 2014 [61]) or political structures (Bebbington 2012 [12]; Orihuela and Thorp 2012 [62]) in the effective management of extractives for equitable and positive benefits. While others, drawing on historical cases, argue that the prevailing capital-intensive models of production cannot ever correlate with social justice (Lederman and Maloney 2007 [54]; Kirsch 2014 [51]).

CSR frameworks for justice—centered around *"buzzwords and fuzzwords"* (Broad 2014) [18] such as "responsible mining," "sustainability," and "social acceptance"—refer to the corporate commitment to doing something "good" for host-communities such as investing in local education and infrastructure. These commitments, in conjunction with self-regulatory

---

[2]Praxis refers to the ways in which a theory or ideia is embodied via exercising of judgment (Smith 1994) [72].

Table 4.4: Key elements of the political economic in effective management of extractives

Effective Management of Extractives (Orihuela and Thorp 2012)
1. The country's ability to extract and use benefits that are derived from a "point source" activity, i.e., the ability to tax and then redistribute wealth;
2. The country's ability to balance short- and medium-term income incentives against long-term development initiatives;
3. The capacity to manage a broad-based, competitive macro-economy;
4. The capacity and will to diversify the macro-economy beyond mining;
5. Effective management of the country's currency through peaks and troughs of demand for mineral exports;
6. The ability and will to monitor and mitigate environmental impacts; and
7. The ability to monitor and mediate internal conflict.

frameworks, reflect the dominant discourse for sustainable mining and social justice (Australian Government 2011 [8]; Newmont Mining Corporation 2014 [61]). Typically, this does not involve change in the mining process itself, but rather, aims to offset or "correct" any injustices *i.e., unequal basic rights or liberties*, that result from exploitation activities. Sustainable Minerals Institute Professor and former Rio Tinto Global Practice Leader, Bruce Harvey, backs these frameworks for development and "ultimate benefits." Additionally, Harvey argues, "net positive impacts" reporting will create trust and forge the necessary partnerships for future success (Harvey 2013) [48].

The case for mining bringing "ultimate benefits" to the communities is a somewhat "utilitarian" perspective in that focus is afforded to the overall maximizing of utility [benefits] rather than its distributive nature. As a result, CSR focused on "giving back" may justify adverse costs incurred by some communities where they are offset by larger, overall benefits. Anthropologist Anthony Bebbington highlights these distributive and redistributive injustices within a political economy, particularly where mines have a tendency to manifest as "political enclaves" or isolated instances of "economic growth" (Armstrong et al. 2014) [5]. Orihuela and Thorp (2012) [62] describe this constrained distribution of benefits as inherent to mining and self-perpetuating:

> *"the very nature of extractives both generates and sustains interpersonal and interregional inequality, with huge implications for political choices and structures"* (Orihuela and Thorp 2012, p. 26) [62].

In order to effectively manage the impacts of mining and ensure socially just outcomes, Orihuela and Thorp (2012) [62] identify seven key elements of the political economy (see Table 4.4).

These mechanisms, Orihuela and Thorp (2012) [62] argue, are central to managing corporate activities and translating extractives into positive benefits for communities.

Lederman and Maloney (2012) [54] offer a more pessimistic view on mining and social justice, describing natural resources as *"cursed goods,"* particularly in developing regions such as Latin America. In this context, the tendency for mechanized production over human capital accumulation hinders a nation's ability to harness and redistribute benefits. This has resulted in lower than expected "growth," decreased long-term job creation and constrained multiplier effects with significant spillovers (Gylfason 2001) [47].

Kirsch (2014) [51] further questions the ability of capitalizt extractives to bring benefits where corporations have neither the structures nor the incentive to achieve equitable and positive impacts. Examining historical cases, Kirsch (2014) [51] highlights that corporations are invariably economically motivated and have the ability to: ignore, refute, or appropriate the terms of critique and scientific data, and; seek to silence or delegitimize their opponents, in order to preserve their bottom line. However, all is not lost—according to Kirsch, exposing of these economically motivated behaviors by civil society is *"one of the primary defenses against the negative consequences of unrestrained capitalizm"* (Kirsch 2014, p. 224) [51]. The rise of NGO critique has been successful in exposing government-corporate relations, changing practices and creating a degree of accountability for the previously unacknowledged "costs" of mining.

4.3   METHODOLOGY

4.3.1   RESEARCH OBJECTIVES

By examining the experiences of indigenous HRDs, this thesis aims to describe some of the unacknowledged yet "real" costs of mining. It is hoped that a deeper understanding of mining in the Latin American geopolitical context may begin to explain why troubling claims of human rights abuse and conflict over territories prevail in the presence of state and corporate policy-based safeguards. Through the use of a gender lens this research provides a more balanced view on the literature, filling gaps related to the voices of marginalized groups such as indigenous to generate holistic understanding of not only the economic, but also the social, gender and environmental impacts of mining.

4.3.2   DATA COLLECTION

The theoretical argument used to justify the qualitative research and methods herein is informed by Case and Light (2011) [21] who highlight the value of using a combination of methodologies to broad the set of questions engineers are able to address.

The primary method of data collection was interviews with WHRDs. The interviews were reflexive and semi-structured, centered around questions of (1) *"what is social justice,"* and (2) *"what are the primary barriers to social justice you are experiencing in the context of mining."* The interviews took place from September 16–17, 2016 during a Forum on Mining and Social Justice

Table 4.5: Country of origin of the interviewers data codes

Abbreviation	Country
04	Guatemala
05	Honduras
02	Peru
01	Ecuador
03	Guatemala

held in Adjuntas, Puerto Rico. The meeting was a joint effort between UWA in collaboration with LAMMP and local organization Casa Pueblo. Table 4.5 indicates the country of origin of the WHRDs interviewed for this study.

The forum was conceived as a capacity building activity for prominent female activists who had experienced gender violence, character defamation and criminalization as a result of their human rights work challenging irresponsible mining policies and practices. By providing a safe environment in which the women could connect and share their experiences, the Latin American activists had the opportunity to reflect on specific difficulties women face when they oppose governmental and corporate decisions which impact on their livelihoods as well as difficulties in accessing justice given their low social status. The interviews were triangulated by ethnographic observations taken during the meeting and archival research to construct a rich textual understanding of the women's experiences.

Ethics approval for the study has been considered under the larger UWA CEME—Engineering Communities and Environmental Critical Mass Research Grant and Ethics Approval. With Guidance from "Qualitative Research Methods: A Data Collector's Field Guide" (Mack, Woodsong, Macqueen, Guest and Namey 2005) [56], written and verbal informed consent for the use of the women's names was obtained.

4.4   DATA ANALYSIS

4.4.1   THEORETICAL FRAMEWORK

Adopting a social constructivist epistemological stance, a theoretical framework was selected to critically analyze the interview transcripts as "a way of knowing" constructed by the women (Baillie 2011 [11]; Willig 2001 [86]). Informed by Rawls' *Justice as Fairness* principle (1973) [67], The Universal Declaration of Human Rights has been utilized to assessing the data against an "*an adequate scheme of equal basic rights and liberties.*" The declaration represents a major internationally binding mechanism and is the cornerstone of current industry-accepted publications such as the *United Nations Guiding Principles (UNGP) on Business and Human Rights* (UNGP, 2011) [80].

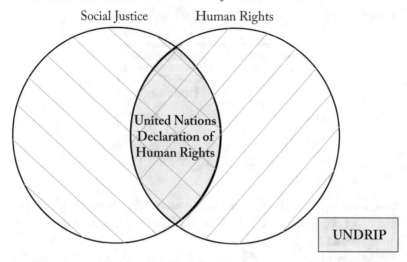

Figure 4.1: Theoretical lens for analyzing the experience of the interviewers.

Additionally, the *UN Declaration on the Rights of Indigenous Peoples* (UNDRIP) (2008) [81] has been drawn upon in assessing indigenous specific rights. Although UN declarations are not typically legally binding, the UN describes UNDRIP as "*the dynamic development of international legal norms…* [that reflects] *the commitment of states to move in certain directions, abiding by certain principles*" (UN 2008, p. 2) [81].

While social justice often articulates a vision that extends beyond the capabilities of human rights mechanisms (Moyn 2015 [59]; Neier 2015 [60]), a rights framework can be seen as a "minimum floor" of requirements to sustain life. At the most basic level, when poverty and social marginalization are the result of discrimination, it is often identifiable as a human rights violation (Chong 2015) [23]. In this way, Chong argues, "*economic and social rights* [outlined in the declaration] *can guide efforts to reform international, state, and corporate actors* [to] *remedy but…not necessarily end social inequity*" (2015 p. 19) [23]. This overlap between social justice and human rights where the Universal Declaration of Human Rights in the space of social justice is read through UNDRIP (Figure 4.1), forms a lens for framing and analyzing the women's experiences (Bernhard and Baillie 2013) [13].

The interview transcripts and field notes, framed through a gender and human rights' lens, were critically analyzed for patterns and themes. Informed by Braun and Clarke (2006) [17], themes were identified as having captured something critical to answering the research question and expressing a level of "patterned response." Identified themes were then enriched by archival research to understand the geopolitical structures underpinning theme.

4.4.2   POTENTIAL LIMITATIONS

All scientific research seeks to understand a given research problem by analyzing technical data systematically through a predefined set of procedures or theoretical "lenses" (Mack et al. 2005) [56]. It is limited in that the theory informing the research and construction of the critical "lens" is ever-changing.

4.4.3   RESULTS AND DISCUSSION

Thematic analysis identified four over-arching themes related to the experiences of the women. These are discussed in what would be a chronological order of social injustice, manifesting from various breaches in human rights articles:

- lack of consultation;

- vilification of activists and criminalization of protest;

- inequality before the law;

  - company protection by private security

  - private security agreements

  - the public sector as a private security client

  - declared "state of emergencies"

- violence, torture, abuse, and murder;

  - use of disproportionate, excessive force

  - kidnap and torture

  - physical abuse, threats, and assassination

Additionally, key take-home messages identified during the meeting at Casa Pueblo are discussed:

- science to strengthen the fight,

- engineering community engagement, and

- grassroots campaigns and the Andean cosmovision.

4.4.4   LACK OF CONSULTATION

Before projects with a significant environmental impact are granted or refused clearance, most Latin American countries require mining corporations to commission independent, impartial EIA (Sánchez-Triana and Enríquez 2007, p. 2) [71]. As part of a study of the potential social and cultural impact of a project both authors consider that the EIA process should include some form of "public consultation" (Sanchez-Triana and Enriquez 2007, p. 3) [71]. However, in practice, companies see the EIA essentially as a tool that helps to foresee and plan for potential environmental problems which if ignored may require expensive changes to the design of the project (Sanchez-Triana and Enriquez 2007, p. 8) [71]. Therefore, the degree to which the concerns of all interested and affected members of the community can influence the scope, design and implementation of the mining project is limited. This imbalance has contributed to public perception of consultations as a suspicious, controversial tool which lacks credibility (Armstrong et al. 2014, p. 41 [5]; BNamericas 2013 [16]). Among the women's group interviewed for this research, public consultations were criticized for becoming a public relations exercise used to notify communities of decisions that had already been taken. There is widespread agreement that any public consultations carried out before the approval of a mining development project should be underpinned by free, prior informed consent (FPIC) which is considered an example of good practice as it is based on the principle that "all people have the right to self-determination." Although it has the backing of the UNDRIP and ILO Convention 169, indigenous WHRDs complain that mining companies do not consider FPIC as an essential requirement of the EIA and remain reluctant to comply with the use of FPIC as a best practice standard (Eye on Latin America 2013) [34].

Article 32.2 of The UN Declaration on the Rights of Indigenous People offers the following guidance regarding consultation for projects impacting indigenous territories:

*"States shall consult and cooperate in good faith with the indigenous peoples concerned, through their own representative institutions in order to obtain their **free** and **informed consent prior** to the approval of any project affecting their lands or territories and other resources, particularly in connection with the development, utilization or exploitation of mineral, water or other resources"* (UN 2008, p. 12) [81].

Given its importance, consultation—perceived as a critical step providing the community with objective and impartial information before consent is granted—was an immediate, common theme emerging from the interviews:

"These projects go against communities' rights and nobody has been consulted" (01);

"the government had given the corporation a slice of the cake without consulting us" (02).

Further, informed consent requires free flow of information about both the positive and negative impacts of a project (UN 2013) [82]. However, the activists interviewed described their experience of the consultation process as "very general and focused on benefits" (02).

In Peru, neoliberal reform, initiated by Alberto Fujimori in 1990 resulted in significant spatial reorganization and changes in control of extractives (Bury 2005) [19]. Privatization of state-owned mineral interests involved the parcelling of communally managed land resources for private sale. At the time, formal transactions were scarce (Castillo 2015) [22], with only 60% of private land holdings in post-agrarian communities recognized by the Peruvian state

(Bury 2005) [19]. One of the consequences of this process was that mining investment often superimposed communally managed lands not formally recognized as such.

Today, indigenous and peasant peoples in regions of the Peruvian highlands, such as the case of Máxima Acuña de Chaupe, continue to fight for recognition of their property rights where mining projects are proposed. Máxima was only recently acquitted of charges for "illegal squatting" on her farmland which was located within a large plot previously acquired by Newmont Mining Corporation for the development of the Conga mine—an expansion of its Yanacocha mine. Newmont Mining never recognized the family's land title, and as a result Máxima suffered years of violence at the hands of armed security forces who tried to eject her from her land (LAMMP 2017) [53].

Land tenure is further clouded by laws distinguishing rights of the surface from those of the subsurface. As in most Latin American countries, Peruvian law establishes that subsurface minerals are property of the state and as such, private actors may be granted concessions underneath individual or communal properties with recognized surface rights (individual or communal) (Del Castillo and Castillo 2003) [30].

Holders of concessions are then required to negotiate with owner(s) of the surface and obtain their consent to purchase their land before the mining development is allowed to proceed (Del Castillo and Castillo 2003) [30].

It has been noted that the company's process of buying land from its owners divides the community into those who agree and those who refuse to sell their land, thus effectively becoming an obstacle to the company (Deonandan and Dougherty 2016) [31]. This "divide and conquer" approach has been perceived as a means by which *governments can wash their hands* (02) and has served to polarize the communities by:

> "[allowing] *the community to generate an internal kind of war between those against and those in favor of selling the land*" (02).

Second, state and corporations traditionally opposed implementation of the complex FPIC consultation process in Peru—in particular to communities which were not included in the official database of indigenous communities published by the Ministry of Culture (Sanborn et al. 2016) [70]. The database made the right to consultation increasingly difficult for peasants communities located in coastal areas as well as the Andean (DPLF 2015) [32] where a high number of mining projects are located. Estimated in thousands and predominantly inhabited by people who used indigenous languages and maintained traditional costumes including collective ownership of the land they occupied (Sanborn et al. 2016) [70] these communities were not included in the database. UN-ECLAC (2014) [78] considers that a rapid urbanization process led to the reclassification of indigenous communities located in the periphery of the most important Latin American cities and mentions Peru as an example where more than 70% of the total indigenous population is concentrated in the Sierra region (214: 57). Sanborn (2016) [70] sees the Agrarian Reform of the 1970s as a pivotal moment when "indigenous communities" became "campesinos" communities. Griffiths (2004) [45] adds that this lack of understanding of how indigenous people were assimilated into the mainstream culture of their countries represents a major obstacle to the recognition of their land rights and their entitlement to FPIC. (Armstrong et al. 2014, p. 41) [5]. Agrarian reform refers to state-led land reform programmes where redistribution of land rights and measures such as access to credit and inputs of production aimed to increase farming productivity. Largely implemented without consideration of indigenous tenure regimes, the reform left families with relatively small parcels of land that would be further subdivided as a result of population growth and more often than not, without legal title to land holdings (Griffiths 2004) [45].

In Peru, President Ollanta Humala stated:

"in the highlands there are agrarian communities… indigenous communities are mostly in the jungle"

when describing protected areas of the nation (Taj and Cespedes 2012) [73].

4.4.5   VILIFICATION OF ACTIVISTS AND CRIMINALIZATION OR PROTEST

Among the interviewees there was a broad perception that the state and corporations are engaging in public smear campaigns to delegitimize protest and vilifying activists:

"I have suffered several attacks, threats, defamation, persecution, and criminalization by the government" (04).

Vilification and criminalization of activists has been identified as a violation of UN Human Rights Article 19, which states:

"Everyone has the right to freedom of opinion and expression… without interference and to seek, receive and impart information and ideas through any media regardless of frontiers" (UN, 1948) [79].

Key political figures such as Former Peruvian Prime Minster Oscar Valdes have labeled those opposing mining projects as *"anti-investment," "anti-development,"* and *"opposed to national interest"* (Poole and Renique 2012) [66] in relentless media campaigns, openly proclaiming: *"Peruvians need investment to create more jobs. What we don't need is disorde"* (Poole and Renique 2012, p. 5) [66].

Similar sentiments are voiced through much of the mainstream media across Latin America. Last year a major Guatemalan newspaper, *Prensa Libre*, published a full-page advertisement describing the actions of human rights organizations as "terrorism"—the campaign was commissioned by a senior representative of one of the country's mining companies (Amnesty International 2016) [2]. Public slander and personifying opponents to mining as a *"minority of extreme violence"* (Vásquez 2013, p. 2) [83], has compromised livelihood and ostracized activists.

Interviewee 04 for this research described how her defamation (*"before people thought of me as a teacher, and now they call me a terrorist"*) has resulted in loss of employment.

Criminalization of activists is also widely reported and perceived as a practice that seeks to silence the protest: *"faced with our claims and our disagreement, the government responded with repression, persecution, criminalization, especially judicial persecution"* (03).

By applying "broad" definitions of the penal code, court proceedings have been instigated against those exercising their right to "assembly" and "association" (Peru Support Group 2013 [13]; UN 1948 [79]). Terms such as "public intimidation," "incitement to violence," "extortion," and "coercion" may be applied to legitimate protest, exposing activists to arbitrary interpretation by judiciary hearing (Aitken 2020) [1].

All the interviewees reported experiences of repression and criminal proceedings related to their protest against mining activities on their territories, describing significant *"pain as a result of very serious charges and a relentless process of harassment and repression"* (02). To this end, impeding one's right to freedom of expression via public smear campaigns, slander and criminalization of protest is far from innocuous. It has served as a tool for discrediting the voices of the women and polarizing the communities in which they live. Further, the fear of becoming embroiled in criminal proceedings has deterred some from participating in what would be legitimate political demonstrations. Dalang (2010) [57] also identified political vilification as a gateway to further abuses of human rights such as *"arbitrary arrest, extrajudicial killing, enforced disappearance, political arrest and detention"* (Umil 2012) [77].

4.5   INEQUITY BEFORE THE LAW

4.5.1   COMPANY PROTECTION BY PRIVATE SECURITY

In response to the need for additional security beyond that provided by host governments, various decrees across Latin America permit companies to commission independent agents and/or engage police to carry out "private security" (Grufides 2013 [46]; Perret 2012 [64]). Private security employees are required to remain impartial and respect human rights in the protection of company assets and personnel (Voluntary Principles 2000) [85]. However, feelings of injustice regarding this impartiality were described by the interviewees, who claim:

> "[there is] *no instance of the judicial power doing anything in our defense* [regarding] *abuse of human rights of defenders*" (01).

4.5.2   PRIVATE SECURITY AGREEMENTS

Throughout the last decade, there has been a steady growth in the private security sector in Latin America with the mining sector employing some 17% of all private security forces (DCAF and UNLIREC 2016) [29]. Privatization of security has resulted in a non-state actor permitted to use force where once only states were allowed to do so—a situation, the Voluntary Principles (2000) [85] highlights, where there is opportunity for abuse of human rights to occur. In relation to this problem, the interviewees described:

"companies bring their own security people and soon they start paying other people to in-timidate and harass us" (05).

Considering the rural context where mining projects take place, of these arrangements, where as much as 50% of the population are living below the poverty line (World Bank Group 2017) [88], a job opportunity with a mining company represents unimaginable benefits in exchange for protection against activists that are publicly portrayed as "terrorists" and "enemies of the state" (Grufides 2016) [88].

4.5.3   THE PUBLIC SECTOR AS A PRIVATE SECURITY CLIENT

In Ecuador and Peru, companies are permitted to hire private and/or public security agents on a "permanent or occasional basis" (Aitken 2020 [1]; Grufides 2013 [46]). When hiring public security officers, companies compensate both the employee and the National Police Force for costs incurred in the protection of their assets. Payment is a: *"…in providing extraordinary additional services"* (Grufides 2013, p. 9) [46].

The fact that armed logistical support provided by mining companies to their security personnel as well as the uniforms used by the private sector are similar in appearance to those used by public forces have caused confusion which often makes it difficult to differentiate between the actions of police and private security forces. This has resulted in a widespread perception that *"the one who receives protection from the state… is the one who pays"* (02). Discriminatory protection by private security and perceived inequality before the law violates Human Rights Article 7: *"All are equal before the law and are entitled without discrimination to equal protection of the law"* (UN 1948) [79].

4.5.4   DECLARED STATE OF EMERGENCIES

In response to national internal or external threats, "state of emergency" may be declared, during which civil liberties are suspended, legislative processes are accelerated and military is deployed. Although described as a last resort for "restoring order," the women interviewed claim: *"The most difficult development we are living is militarization of communities"* (05). Militarization of zones where opposition to mining is intense is associated with heightened police and private presence, security, and military presence resulting in increased criminal behaviors, drug use and perpetuating further human rights breaches including assault and sexual abuse (Martinez 2010) [57].

From 1995–2015, Peru, Guatemala, and Ecuador reported more manmade "state of emergencies" than any other nation globally (Table 4.2) (Zwitter, Fister, and Groeneweg 2016) [91]. Further, manmade disasters were associated with prolonged duration of state of emergency in regions demanding rights vis-a-vis mining development and correlated with increased reports of human rights abuses during this time including: arbitrary arrest (Article 9), right to privacy (Article 12), and right to freedom of association (Article 19) (Zwitter et al. 2016) [91].

Declared states of emergencies are perceived as *"a way of silencing us* [activists] *…a means for the government to guarantee investors that their project will go ahead"* (05). The prolonged presence of military in regions and suspension of human rights such as freedom of 'assembly' without impunity have rendered activists unable to mobilize in defense of their territories.

4.5.5   VIOLENCE

The culmination of these human rights abuses has been an alarming trend of excessive police force, kidnappings, torture, abuse and murder in the context of mining (LAMMP 2017 [53]; Frontline Defenders 2017 [38]). The Forum in Puerto Rico provided a safe and secure platform for the women to share their personal experiences where the common consensus was *"we have been victims of persecution, physical, and psychological abuse"* (05).

In light of Article 5 of the UN Human Rights Declaration: *"No-one shall be subjected to torture or to cruel, inhumane or degrading treatment or punishment,"* these instances represent a clear violation of basic liberties. While it should be acknowledged that there is, in some cases, evidence suggesting both state authorities and HRDS have acted unlawfully, the focus herein is the construction and understanding of these experiences by WHRDs.

Interview transcripts and field notes referring to three particular cases were later triangulated by archival research. This has enabled a deeper understanding of **who** the parties involved are?, **what** is known about the chain of events surrounding the experiences?, and **how** they are occurring in the mining context? These three events will be discussed in turn.

4.5.6   USE OF DISPROPORTIONATE, EXCESSIVE FORCE

When responding to conflict, host-states are obligated to ensure that policing is in line with global standards for human rights; use of force, and; firearms (Amnesty International 2014) [2]. Central principles to both ordinary and extraordinary (i.e., state of emergency) public policing include:

- use of force when "strictly necessary" and only to the extent required to perform their duty, and;

- use of force in proportion to the seriousness of the offence so as to minimize injury and respect human life (Voluntary Principles 2000) [85].

Yet, the police response to a peaceful protest in October, 2015 in Ecuador, is questioned by one of the WHRD interviewed:

"On this occasion we were attacked by the police, we were beaten, we were swept aside, a group of women—many of them elderly, close to 80 years old, were assaulted in this way and were arbitrarily arrested" (01).

The violent incident took place when indigenous HRDs peacefully protesting the controversial Rio Blanco Mine at the project's inauguration ceremony, attended by Ecuadorian President Rafael Correa. The women were displaying a sign that read *"Minería Responsable Cuento*

Miserable" (Responsible Mining a miserable tale story) when they were allegedly beaten and arbitrarily detained by the police for the duration of the ceremony (Frontline Defenders 2016) [37].

On July 8, 2016, an Ecuadorian court ruled against the women regarding a complaint they submitted for the mistreatment by police. In the hearing:

> "*we provided evidence (including documents and film) of what had happened to us that day when the police attacked us,* [despite this] *the Ombudsman accepted what the police told them, that a group of elderly women had attacked them… they had been victims of vicious attacks and aggression. After watching the film showing how we were dragged by our hair, the Ombudsman accepted the police argument and rejected our request for justice*" (01).

This exemplifies a clear case of excessive force and arbitrary arrest where the unarmed women, some elderly, were controlled and detained pursuant to "*maintaining integrity and security of the participations* [of the inauguration]" (The Ombudsman in Azuay—Zone 6, quoted in FrontLine Defenders, 2016) [37]. After what has been a "*terrible and demoralizing*" experience, the women are left "*totally unprotected*" with the feeling that "*there is no justice for us women, there is no justice for communities*" (01).

4.5.7   KIDNAP AND TORTURE

In Peru, protest against the Majaz mine (owned by S. A. Majaz, a subsidiary of UK-based Moneterrico Metals Plc) resulted in other forms of abuse:

> "*In August 2005, during a peaceful five-day march, … 28 protesters (including two women) were kidnapped …by Rio Blanco security forces and held captive for three days. Being the only women in the group, they faced gendered abuse, kept half-naked in a small toilet with plastic black sacks over their heads, their feet and hands tied. Both women recall receiving rape threats and constant sexual harassment*" (Wong and Rowe 2014, p. 3 [89]).

One of the kidnapped women described her harrowing experience "*as some of the worst of my life. When I was beaten it changed my world… it was as though a tornado had destroyed everything*" (Wong and Rowe 2014, p. 4) [89].

Adding insult to injury, the Provincial (Huancabamba) Attorney charged the 28 victimis with participating in illegal protest—this case was later dropped due to lack of sufficient evidence (Wong and Rowe 2014) [89]. An anonymous "tip-off" in the form of photographic evidence of the beaten activists, received by LAMMP Project Director Glevys Rondon, alerted the organization but it was the London-based law firm Leigh Day which undertook the mammoth task of providing the evidence necessary to pursue the (then) owner of Monterrico Metals Plc for the abuses. Although the company never admitted liability, in 2011 all 28 protesters held captive by Monterrico Metals Plc received financial compensation in an out-of-court settlement.

ULAM's July 2014 submission to the UN commission on the status of the women details the ways in which both women as well as other compensated captives continue to suffer harassment, intimidation and death threats. A timeline spanning 2005 until publication in 2014 details text messages, assaults and home break-ins to the effect of: "*shameless robber, return the money or we will beat you and kill you*" (Wong and Rowe 2014) [89]. The women have suffered additional gender discriminations and been left isolated by community members who "*blame them for sexual abuse because they took part in demonstration*" and label them as "*dirty women*" (02).

These ongoing abuses have not been investigated by police and as a result the women live in constant fear: "*afraid to leave their homes*" (Wong and Rowe 2014) [89]. Unfortunately, what happened to the protesters against the Majaz project is not isolated: in total 53 people have been killed and almost 1,500 injured in social conflicts across Peru, relating to mining activities since 2011 (The Economist 2016) [74].

4.5.8   PHYSICAL ABUSE, THREATS, AND ASSASSINATION

Violation of of UNDRIP Article 5 was without a doubt the most discussed issue during the Forum. In particular, the recent assassination of prominent activist Berta Cáceres shot dead in her La Esperanza home on March 3, 2016—a victim of a 10-year campaign against construction of the Agua Zarca Dam, a project that would cut off a significant water source for the downstream Lenca people. Berta, the 2015 recipient of the Environmental Prize, was renowned for her brave and tireless efforts amidst threats on her life and assassination of fellow activists protesting the dam.

"*They killed … …. Berta Cáceres because they want to silence us*" (05).

In spite of a police investigation, determining the "they" in this context is complex. Five men have been arrested over the murder, including one who is an active major in the armed forces and head of security for the Agua Zarca Dam. "*The people they threw in jail are not the real killers… [Berta] was killed by the government, the transnational corporations, the banks that finance these projects*" (05).

This perception resonates globally, where the investigation of Berta's murder has been condemned, prompting an independent inquiry "*to try to uncover the intellectual authors behind the assassination of Cáceres*" (Lakhani 2016) [52]. The inquiry is currently investigating: the state's response to the murder; the potential role of international companies; and the international banks that supported the project (Lakhani 2016) [52]. The statement of former Honduran Attorney General that "*public prosecution is only able to investigate 20% of killings due to insufficient resources*" reinforces this need for an international inquiry (the Attorney General was removed by the Congress after making this statement).

Unfortunately, the murder of Berta Cáceres is not isolated—in 2017 Global Witness declared Honduras "the deadliest place to defend the planet" where 123 activists have been murdered since the military coup in 2009 (Global Witness 2017) [40]. Across Latin America, activists are facing a similar fate: in 2015 65% (122 out of 185) of globally registered murders of activists—defending land, territory, or the environment—occurred in Latin America (Oxfam 2016) [35].

4.5.9   LEARNING FROM CASA PUEBLO

Casa Pueblo was selected to host the Forum as it exemplifies a positive, socially just outcome for rural and indigenous communities threatened by irresponsible mining practices. Key messages were derived from the meeting that can assist not only activists and HRDs, but also engineers in better understanding their potential role in communities.

4.5.10   SCIENCE TO STRENGTHEN THE FIGHT

Adjuntas local and civil engineer, Alexis Massol-González, founded Casa Pueblo in 1980 in response to a proposal by the Puerto Rican government for the extraction of 17 deposits of silver, gold, and copper. The deposits underlie fragile rainforest responsible for water supply to no less than 1.5 million people on the island of Puerto Rico (Goldman Environmental Foundation 2017) [42]. Alexis conducted research that showed mining would compromise water supply and engaged with the Adjuntas community, educating them about how mining would impact their livelihood. The product of a 15-year fight was the 1985 overturning of the mining proposal and establishment of a permanent organization for protecting natural, cultural and human resources (Casa Pueblo 2016) [20].

At Casa Pueblo, one of the speakers reflected on realizing the value of science through her own experiences:

> *"I have been one of those activist who went on marches… but as time went by we came to realize that apart from being at the front of the protest with your body, we also need scientific tools that allow us to put in the bin all the arguments miners throw at us… we have to provide communities with evidence of the consequences of mining on people and the environment"* (02).

4.5.11   ENGINEERING COMMUNITY ENGAGEMENT

Alexis' work exemplifies the way engineers can support social justice for vulnerable communities. Within the modern mining paradigm, the role of an engineer calls for little contact with communities throughout scoping and impact assessment stages (Kemp and Owen 2013) [49].

One of the participants at the Forum described her experience of working with mining students:

> *"…what the universities teach you is cold. The student is expected to have scientific knowledge of where the mineral is located and what are the best ways to bring it out, but they don't fully understand the consequences of their actions"* (02).

This perceived *"lack of humanity"* (02) may be a product of the corporate "production model" and the ingrained division of labor it constructs (Armstrong et al. 2014) [5]. Mining engineers are typically hired to solve problems, plan and operate engineering aspects of extractives (AusIMM n.d.) [7]. The potential impacts and externalities are reduced to "data inputs" for design, obtained in the field by sociologists, anthropologists, or those qualified in the area of interest (Armstrong et al. 2014) [5].

Furthermore, where projects are located in remote regions, it is common for engineers to reside in self-sufficient "enclaves" developed specifically for mining staff (Bebbington 2012) [12], or commute to work on a FIFO basis. As a result, this anonymity between engineers and the communities in which they operate continues from scoping through to production phase of mining and may prevent engineers from ever truly observing the impacts of their work. During a gathering of WHRDs supported by ULAM Máxima Acuña recalled her attempts to inform engineers of HR abuses committed by personnel of Newmont Mining. Maxima reported that her attempts were not only ignored but dismissed without any formal inquiry by the engineer in charge (ULAM: 2014 "Submission to the UN Commission on the status of women" [76]; Armstrong et al. (2014 p. 48 [5]).

Here, it is evident that Franklin's warning of the potential for dissociation of engineers from externalities (and those who receive their technology) under the "production model" has manifested in the Latin America context (Franklin 1999) [36].

In a moment of reflection with the interviewees, the notion of a new type of engineer, a "community engineer," was raised. This role, it was proposed, would aim to bridge the gap between CSR frameworks and CSR praxis in the context of mining. In this way, the modern engineer may begin to look beyond those impacts measured by the "audit culture" framework and begin to holistically understand what their work represents for communities. Further, it was acknowledged that the contribution engineers and scientists offer is unique and can "*contribute ideas about how to develop other alternatives or possibilities of living in harmony with nature*" (02).

4.5.12   GRASSROOTS CAMPAIGNS AND THE ANDEAN COSMOVISION

Understanding the opposition to mining communicated at the Casa Pueblo event requires comprehension of the indigenous "way of knowing" and perception of the world. An essential component of the anti-mining protests are concerns for the well-being of "mother earth" as well as respect for the Andean Cosmovision. The Andean cosmovision is the indigenous culture of the high Andes and has its inception in the Pre-Columbian era subsistence farming societies. It is a way of perceiving and interacting with the world rather than a distinct set of concepts or beliefs (Gordon 2016) [43].

A significant point of difference between Eurocentric-Columbian and Andean epistemologies is the mode of perception they manifest (Ardeleanu 2014) [4]. Under the cosmovision, a peaceful symbiosis with *pachamama* (mother earth) and nature informs farming practices and deeper identity of the people:

"as ancestral people we have a very special relationship with mother earth. If you contaminate the earth's blood, you are contaminating our blood as well—it is part of ourselves and our identity" (05).

From this stance, we can begin to understand how the notion of mining and "pillaging" the subsurface fundamentally conflicts with the ideals of the cosmovision:

We do not conceive the idea of exploitation or exploration, because in our imagination, and that of my grandmothers and grandfathers, the large deposits of silver and gold are like the bones of the earth" (03).

The changes in the political economy and dislocation of indigenous peoples from their land (i.e., identity) that support mining, have been likened to a modern colonialism where: *"…they are trying to erase our cosmovision"* (05).

The use of a gender lens makes evident that the extent of the impact of mining activities in their lives is intimately linked to their multiple responsibilities within the family unit not only as careers for their children but also for general duties of farming and looking after livestock

"When water is contaminated we have to work harder to subsist" (05);

"We are … the first to react because we are the first to feel the impact, the first that can sit down and admire nature because of our direct contact with it" (02).

Despite this, consultation over land tenure is "men's business;" and mining is "men's" work (Wright 2017) [87]. The value of grassroots campaigns in educating communities and "claiming back rights" is exemplified by Casa Pueblo who continues to engage the Adjuntas community and forge relationships between people and the earth. One of the speakers at the forum described similar success in her local area where she helped organize a good faith consultation: *"more than 27,000 people including women; children; men and the elderly attended and said no to mining—not because of the* [lack of] *royalties—the connection is not economic, the connection is about life"* (ALC in Frontline Defenders 2014) [39].

"With the men …the ones making decisions to sell off the land—we had to make an effort to bring about a change in their perception of land tenure … the bones of our grandmothers are in that soil, so how dare you sell that land!" (04).

4.6   SUMMARY

The discussion herein has presented evidence in clear breaches of human rights abuse in the context of mining, from initial scoping stages where EIA has failed to consult key stakeholders (UNDRIP Article 32.2) such as women, traditionally excluded from consultations of development projects (UNDRIP Article 19), through extraction and production stages where influxes of security forces have resulted in violence, abuse, and even murder of activists (UNDRIP Articles 7 and 5). Abuses and violations of human rights have taken place despite the presence of global standards such as the UN Guiding Principles which provide a framework regarding the state *duty* to protect and corporate responsibility to respect human rights. To this end, it is clear that themes of social justice praxis identified in the literature do not necessarily equate with "a fully adequate scheme of equal basic rights and liberties for all."

Complexities of the geopolitical context including state obligations to protect FDI, vilification of activists, and requirement for private security to supplement policing in the protection of company assets has collectively resulted in the marginalization and silencing of vulnerable groups who so often are the first to notice and suffer the impacts of mining.

Support from local and international NGOs has helped to increase awareness of activists struggle to access remedy and reparation. However, it is clear that no economic compensation can truly equate with justice for these indigenous people where the idea of exploiting mother earth is inconceivable and conflicts with Andean cosmovision.

4.7 CONCLUSIONS

Examining the experiences of the WHRDs from Latin America interviewed for this study, it is evident that human rights abuse and social injustice prevail in the context of mining. The reality of neoliberal reform for the interviewees, who represent some of the most vulnerable groups of society, has been a dangerous cycle of vilification, isolation, threats, kidnap, torture, abuse, sexual abuse, and in some cases murder. While industry and NGOs alike have attempted to provide frameworks for "respect, remedy, and redress," the message clearly communicated herein is that no amount of financial compensation can justify the abuses suffered by vulnerable groups at the hands of mining. At present, it remains the state's "*duty to protect*" and the corporation's "*responsibility to protect*" however, establishment of a legally binding instrument on business and human rights is certain to change the way that mining companies assess the social and environmental impacts of their work.

As engineers, we are equipped with the knowledge and capability to be at the center of such a shift, where engaging with local communities and providing evidence of the real costs of mining, for all those affected, can began to slow what is a dangerous and self-perpetuating chain of human rights abuse.

4.8 NEWS ARTICLES AND OTHER PUBLICATIONS

[1] Aitken, J. (2020). *Chapter 3 of This Publication*. 99, 100

[2] Amnesty International (2014). *Mining in Guatemala: Rights at Risk*. London, UK, Amnesty International Publications. http://www.amnesty.ca/sites/amnesty/files/mining-in-guatemala-rights-at-risk-eng.pdf 98, 101

[3] Amnesty International (2016). Honduras/Guatemala: Attacks on the rise in world's deadliest countries for environmental activists. *Amnesty International*.

https://www.amnesty.org/en/latest/news/2016/09/honduras-guatemala-ataques-en-aumento-en-los-paises-mas-mortiferos-del-mundo-para-los-activistas-ambientales/

[4] Ardeleanu, C. (2014). An analysis of an Andean cosmovision: Nature, culture, ecology, and cosmos. Masters dissertation. http://scholarworks.sjsu.edu/cgi/viewcontent.cgi?article=8037&context=etd_theses 105

[5] Armstrong, R., Baillie, C., and Cumming-Potvin, W. (2014). *Mining and Communities: Understanding the Context of Engineering Practice.* Synthesis Lectures on Engineering, Technology, and Society. San Rafael, CA, Morgan & Claypool Publishers. http://dx.doi.org/10.2200/S00564ED1V01Y201401ETS021 DOI: 10.2200/s00564ed1v01y201401ets021. 85, 91, 96, 98, 104, 105

[6] Armstrong, R., Baillie, C., Fourie, A., and Rondon, G. (2014). Mining and community engagement in Peru: Communities telling their stories to inform future practice. *International Mining for Development Centre*, University of Western Australia. https://im4dc.org/wp-content/uploads/2014/09/Mining-and-community-engagement-in-Peru-Complete-Report.pdf

[7] AusIMM (n.d.). Minerals industry Careers. Rich in Discovery. *Mining Engineering.* https://www.ausimm.com.au/content/docs/mining_engineering_brochure.pdf 105

[8] Australian Government. Department of Industry, Innovation and Science (2011). *Leading Practice Sustainable Development Program for the Mining Industry.* https://industry.gov.au/resource/Documents/LPSDP/guideLPSD.pdf 90, 91

[9] Australian Human Rights Commission (2008). *Corporate Social Responsibility and Human Rights.* https://www.humanrights.gov.au/publications/corporate-social-responsibility-human-rights 86

[10] Baillie, C. and Douglas, E. (2014). Confusions and conventions: Qualitative research in engineering education. *Journal of Engineering Education*, 103(1):1–7. http://dx.doi.org/10.1002/jee.20031 DOI: 10.1002/jee.20031.

[11] Baillie, C. (2011). A guide to the novice engineering education researcher. *Journal of Engineering Education*, 10(1):100. 93

[12] Bebbington, A. (Ed.) (2012). *Social Conflict, Economic Development and Extractive Industry.* New York, Routledge. 85, 86, 87, 90, 105

[13] Bernhard, J. and Baillie, B. (2013). Standards for quality of research. *Proc. of the Research in Engineering Education Symposium.* Kuala Lumpur. DOI: 10.2139/ssrn.2562760. 94, 99

[14] Bilchitz, D. (2014). The necessity for a business and human rights treaty. University of Johannesburg. http:/dx.doi.org/10.2139/ssrn.2562760

[15] BNamericas (2011). Codelco tech revamp to make low-grade copper ore mining profitable. Last accessed February 20, 2019. http://www.bnamericas.com/en/news/ict/codelco-tech-revamp-to-make-low-grade-copper-ore-mining-profitable 85

[16] BNamericas (2013). UN officials warn of rise in mining conflicts in indigenous communities. Last accessed February 20, 2019. https://www.bnamericas.com/en/news/miningandmetals/un-officials-warn-of-rise-in-mining-conflicts-in-indigenous-communities 96

[17] Braun, V. and Clarke, V. (2006). Using thematic analysis in psychology. *Qualitative Research in Psychology*, 3(2):77–101. http://dx.doi.org/10.1191/1478088706qp063oa DOI: 10.1191/1478088706qp063oa. 94

[18] Broad, R. (2014). Responsible mining: Moving from a buzzword to real responsibility. *The Extractive Industries and Society 1*, (1):4–6. http://dx.doi.org/10.1016/j.exis.2014.01.001 DOI: 10.1016/j.exis.2014.01.001. 85, 90

[19] Bury, J. (2005). Mining mountains: Neoliberalism, land tenure, livelihoods and the new Peruvian industry in Cajamarca. *Environment and Planning*, 37:221–239. http://dx.doi.org/10.1068/a371 DOI: 10.1068/a371. 86, 96, 97

[20] Casa Pueblo (2016). *House Pueblo.* http://casapueblo.org/index.php/proyectos/casa-pueblo/ 104

[21] Case, J. M. and Light, G. (2011). Emerging methodologies in engineering education research. *Journal of Engineering Education 100*, (1):186–210. https://search.proquest.com/docview/857255419?accountid=14681 92

[22] Castillo, A. M. (2015). Transforming Andean space: Local experiences of mining development in Peru. Doctoral dissertation, University of Queensland, Australia. DOI: 10.14264/uql.2015.958. 96

[23] Chong, D. (2015). How human rights can address socioeconomic inequality. In D. Lettinga and L. van Troost (Eds.), *Can Human Rights Bring Justice?*, pages 19–26, Amsterdam, Netherlands, Amnesty International Netherlands. 94

[24] Council on Hemispheric Affairs (2014). Canadian mining in Latin America: Exploitation, inconsistency, and neglect. Last accessed February 20, 2019. http://www.coha.org/canadian-mining-in-latin-america-exploitation-inconsistency-and-neglect/ 86

[25] Costa Rica News. https://thecostaricanews.com/open-pit-mining-prohibited-in-costa-rica/

[26] Cramme, O. and Diamond, P. (Eds.) (2009). *Social Justice in a Global Age*. Cambridge, UK, Polity Press. 90

[27] Dajer, T. (2015). High in the Andes, a mine eats a 400-year-old city, National Geographic. http://news.nationalgeographic.com/2015/12/151202-Cerro-de-Pasco-Peru-Volcan-mine-eats-city-environment/ 85

[28] Denniss, R. (2014). Miners reveal a poverty of thinking on coal. Published in The Camberra Times. Last accessed February 20, 2019. http://www.canberratimes.com.au/comment/miners-reveal-a-poverty-of-thinking-on-coal-20141107--11il9t.html 87

[29] DCAF and UNLIREC (2016). *Armed Private Security in Latin America and the Caribbean. Oversight and accountability in an evolving context*. United Nations. http://dx.doi.org/10.13140/RG.2.2.14256.15363 99

[30] Del Castillo, L. and Castillo, P. (2003). La servidumbre minera y la propiedad de la tierra agrícola. *Tierras Agrícolas y Servidumbre Minera*, pages 7–56. Lima, Peru, Centro Peruano de Estudios Sociales. 97

[31] Deonandan, K. and Dougherty, M. L. (2016). *Mining in Latin America: Critical Approaches to the New Extraction*. Oxon, Routledge. DOI: 10.4324/9781315686226. 97

[32] DPLF (2015). Derecho a la consulta y al consentimiento previo, libre e informado en América Latina. Avances y desafíos para su implementación en Bolivia, Brasil, Chile, Colombia, Guatemala y Perú. Last accessed on February 25, 2019. 97

[33] DW Akadamie (2017). El Salvador becomes first country to ban metals mining. *Deutsche Well*. http://p.dw.com/p/2aHe4

[34] Eye on Latin America (2013). Colombia: The dark side of its coal mining success. Last accessed February 20, 2019. https://eyeonlatinamerica.com/2013/11/13/colombia-dark-side-coal-mining/ 96

[35] Ferreyra, C. (2016). The risk of defending: The intensification of attacks on human rights activists in Latin America: Prepared for Oxfam International. https://www.oxfam.org/sites/www.oxfam.org/files/bn-el-riesgo-de-defender-251016-es_0.pdf 104

[36] Franklin, U. M. (1999). *The Real World of Technology*. Toronto, ON, House of Anansi Press. 89, 90, 105

[37] Frontline Defenders (2016). *Case History: Lina Solano Ortiz*. https://www.frontlinedefenders.org/en/case/case-history-lina-solano-ortiz 102

[38] Frontline Defenders (2017). Honduran and international march to commemorate Berta Caceres. 101

[39] Frontline Defenders (2014). *Aura Lolita Chavez*. https://www.frontlinedefenders.org/en/profile/aura-lolita-chavez 106

[40] Global Witness (2017). Honduras: The deadliest country in the world for environmental activism. *Global Witness*. https://www.globalwitness.org/en/campaigns/environmental-activists/honduras-deadliest-country-world-environmental-activism/ 103

[41] Global Witness (2016). Defenders of the Earth. Last accessed 20th February 2019. https://www.globalwitness.org/en/campaigns/environmental-activists/defenders-earth/

[42] Goldman Environmental Foundation (2017). Alexis Massol Gonzalez 2002 Goldman Prize Recipient Islands and Island Nations. http://www.goldmanprize.org/recipient/alexis-massol-gonzalez/ 104

[43] Gordon, O. E. (2016). The Andean cosmovision: Connecting to the heart of nature. *International Conference on Positive Psychology and Cognitive Behavioural Therapy*. http://dx.doi.org/10.4172/2161--0487.C1.001 DOI: 10.4172/2161-0487.c1.001. 105

[44] Graulau, J. (2008). Is mining good for development?: The intellectual history of an unsettled question. *Progress in Development Studies 8*, (2):129–162. DOI: 10.1177/146499340700800201. 85

[45] Griffiths, T. (2004). Indigenous peoples, land tenure and land policy in Latin America. *Food and Agriculture Corporate Document Repository*. http://www.fao.org/docrep/007/y5407t/y5407t0a.htm 97, 98

[46] GRUFIDES (2013). Police in the pay of mining companies. http://assets.gfbv.ch/downloads/report_english_def_2_12_13.pdf 99, 100

[47] Gylfason, T. (2001). Natural resources and economic growth: What is the connection? *CESifo Working Paper No. 530*. https://ssrn.com/abstract=279679 DOI: 10.1007/978-3-642-57464-1_5. 92

[48] Harvey, B. (2013). Social development will not deliver social license to operate for the extractive sector. *The Extractive Industries and Society 1*, (1):7–11. http://dx.doi.org/10.1016/j.exis.2013.11.001 DOI: 10.1016/j.exis.2013.11.001. 85, 90, 91

[49] Kemp, D. and Owen, J. R. (2013). Social license and mining: A critical perspective. *Resources Policy 38*, pages 29–35. http://dx.doi.org/10.1016/j.resourpol.2012.06.016 DOI: 10.1016/j.resourpol.2012.06.016. 104

[50] Kemp, D., Owen, J. R., and van de Graaff, S. (2011). Corporate social responsibility, mining andaudit culture. *Journal of Cleaner Production 24*, pages 1–10. http://dw.doi.org/10.1016/j.jclepro.2011.11.002 DOI: 10.1016/j.jclepro.2011.11.002. 86

[51] Kirsch, S. (2014). *Mining Capitalism the Relationship between Corporations and their Critics*. Berkeley, CA, University of California Press. DOI: 10.1525/9780520957596. 90, 92

[52] Lakhani, N. (2016). Berta Caceres murder: International lawyers launch new investigation. *The Guardian.* https://www.theguardian.com/world/2016/nov/15/berta-caceres-murder-honduras-international-investigation 103

[53] LAMMP (2017). *Máxima Acuña de Chaupe.* http://lammp.org/campaigns/support-for-maxima-acuna/ 85, 97, 101

[54] Lederman, D. and Maloney, W. F. (2012). Does what you export matter? In search of empirical guidance for industrial policies. Washington, DC, International Bank for Reconstruction and Development/The World Bank. http://dx.doi.org/10.1596/978--0-8213-8491-6 90, 92

[55] Lettinga, D. and van Troost, L. (Eds.), (2015). *Can Human Rights Bring Social Justice?*. Amsterdam, Netherlands Amnesty International Netherlands. 89

[56] Mack, N., Woodsong, C., Macqueen, K., Guest, G., and Namey, E. (2005). Qualitative research methods: A data collector's field guide. *Family Health International*, Durham, NC. https://www.learningdomain.com/Chapter2.Methods.QR.doc 93, 95

[57] Martinez, D. T. (2010). Mining and the environment in Peru: A gendered perspective. Honours dissertation, Nottingham Trent University, UK. http://lammp.org/resources/dissertation-woman-mining/ 99, 100

[58] Mill, J. S. (1863). *Utilitarianism*, 1st ed., London, Parker, Son & Bourn, West Strand. 89

[59] Moyn, S. (2015). Human rights and the age of inequality. In D. Lettinga and L. van Troost (Eds.), *Can Human Rights Bring Justice?*, pages 13–18. Amsterdam, Nederland, Amnesty International Netherlands. 94

[60] Neier, A. (2015). Human rights and social justice: Separate causes. In D. Lettinga and L. van Troost (Eds.), *Can Human Rights Bring Justice?*, pages 47–52. Amsterdam, Nederland, Amnesty International Netherlands. 94

[61] Newmont Mining Corporation, (2014). Beyond the mine—our 2014 social and environmental performance. http://sustainabilityreport.newmont.com/2014/_docs/newmont-beyond-the-mine-sustainability-report-2014.pdf 90, 91

[62] Orihuela, J. C. and Thorp, R. (2012). The political economy of managing extractives in Bolivia, Ecuador, and Peru. In A. Bebbington (Ed.), *Social Conflict, Economic Development and Extractive Industry*, pages 27–45. New York, Routledge. 90, 91, 92

[63] Palumbo, G. and Malkin, E. (2017). El Salvador, prizing water over gold, bans all metal mining. *The New York Times*. https://www.nytimes.com/2017/03/29/world/americas/el-salvador-prizing-water-over-gold-bans-all-metal-mining.html?_r=0 86

[64] Perret, A. (2012). Private military and security companies in Latin America: A regional challenge. *Congress of the Latin American Studies Association*, San Francisco, CA. 99

[65] Peru Support Group, (2013). Criminalisation of social protest. http://www.perusupportgroup.org.uk/peru-human-rights-social-protest.htm

[66] Poole, D. and Rénique, G. (2012). Peru: Humala takes off his gloves. *NACLA Report on the Americas 45*, (1):4–5. http://dx.doi.org/10.1080/10714839.2012.11722104 DOI: 10.1080/10714839.2012.11722104. 98

[67] Rawls, J. (1973). *A Theory of Justice*. Oxford, UK, Oxford University Press. 89, 90, 93

[68] Rawls, J. (1999). *A Theory of Justice*. (Revised edition). Cambridge, MA, Belknap Press of Harvard University Press. 89

[69] Richards, P. (2013). *Race and the Chilean Miracle: Neoliberalism, Democracy, and Indigenous Rights*. Pittsburgh, PA, University of Pittsburgh Press. DOI: 10.2307/j.ctt7zw936. 86

[70] Sanborn, C., Hurtado, V., and Ramírez, T. (2016). La consulta previa en el Perú: Avances y retos. Universidad del Pacifico. 97

[71] Sanchez-Triana, E. and Enríquez, S. (2007). A comparative analysis of environmental impact analysis systems in Latin America. *Annual Conference of the International Association for Impact Assessment*. 95, 96

[72] Smith, M. K. (1994). *Local Education. Community, Conversation, Action*. Buckingham: Open University Press. 90

[73] Taj, M. and Cespedes, T. (2012). Exclusive: Peru rolling back indigenous law in win for mining sector. *Reuters*. http://www.reuters.com/article/us-peru-mining-indigenous-idUSBRE9400CG20130501 98

[74] The Economist (2016). Mining in Latin America: From conflict to co-operation. *The Economist*. http://www.economist.com/news/americas/21690100-big-miners-have-better-record-their-critics-claim-it-up-governments-balance 86, 103

[75] The Costa Rica News (2018). Open-pit mining prohibited in Costa Rica. Last accessed February 20, 2019. https://thecostaricanews.com/open-pit-mining-prohibited-in-costa-rica/ 86

[76] ULAM (2014). Unión Latino Americana de Mujeres. Submission to the UN Commission on the status of women. 105

[77] Umil, A. M. (2012). Vilification of activists, insidious form of human rights violation. *Bulatlat*. http://bulatlat.com/main/2012/10/02/vilification-of-activists-insidious-form-of-human-rights-violation/ 99

[78] United Nations-ECLAC (2014). Guaranteeing indigenous people's rights in Latin America. Progress in the past decade and remaining challenges. http://www.fao.org/fileadmin/user_upload/AGRO_Noticias/smart_territories/docs/Guaranteeing_indigenous_LA_summary.pdf 97

[79] United Nations (1948). The universal declaration of human rights. http://www.ohchr.org/EN/UDHR/Documents/UDHR_Translations/eng.pdf 98, 99, 100

[80] United Nations. Office of the High Commissioner, (2011). United Nations guiding principles on business and human rights: Implementing the United Nations protect, respect, remedy, and framework. http://www.ohchr.org/Documents/Publications/GuidingPrinciplesBusinessHR_EN.pdf 93

[81] United Nations (2008). United Nations declaration on the rights of indigenous peoples (UNDRIP). http://www.un.org/esa/socdev/unpfii/documents/DRIPS_en.pdf 94, 96

[82] United Nations. Office of the High Commissioner (2013). Free, prior and informed consent of indigenous peoples. http://www.ohchr.org/Documents/Issues/IPeoples/FreePriorandInformedConsent.pdf 96

[83] Vásquez Chuquilin, M. (2013). La criminalización de la protesta social en el Perú: Un análisis a la luz del caso Conga en Cajamarca. Grupo de formación e intervención para el desarrollo sostenible (GRUFIDES). Cajamarca, Peru. 98

[84] Vittor, L. (2014). Indigenous people and resistance to mining projects, (English version). *Harvard Review of Latin America*. Last accessed on February 25, 2019. https://revista.drclas.harvard.edu/book/indigenous-people-and-resistance-mining-projects 85

[85] Voluntary Principles (2000). Voluntary principles on security and human rights. http://www.voluntaryprinciples.org/what-are-the-voluntary-principles/ 99, 101

[86] Willig, C. (2001). *Introducing Qualitative Research in Psychology: Adventures in Theory and Method*. Buckingham, UK, Open University Press. 93

[87] Wright, C. (2017). Expanding extractive industries, contracting human rights. Emergency powers and natural resource governance in Latin America. *Contracting Human Rights Project*. University of California, Santa Barbara. 106

[88] World Bank Group (2017). World Bank Open Data. http://data.worldbank.org/ 86, 87, 100

[89] Wong, M. and Rowe, D. (2014). Peru: Rights of rural and indigenous women endangered by transnational mining: Prepared to the UN commission on the status of women. http://lammp.org/wp-content/uploads/2014/10/ULAM-submission-to-the-UN-Comission-on-the-status-of-women-July-2014-PERU-3.pdf 102, 103

[90] Zwitter, A. J., Prins, A. E., and Pannwitz, H. (2014). State of emergency mapping database. *University of Groningen Faulty of Law Research Paper Series.* http://dx.doi.org/10.2139/ssrn.2428254 85

[91] Zwitter, A., Fister, L., and Groeneweg, S. (2016). State of emergency mapping project: Database. http://emergencymapping.org/database2.html DOI: 10.1007/978-3-658-16588-8_19. 100

CHAPTER 5

Everyday Gender Violence in Peru in the Context of Extractive Operations

Glevys Rondon

5.1   INTRODUCTION

A study carried out by the World Health Organization (2005) [92] identified high levels of violence in Peru's two major cities: 51% of women in Lima and 69% in Cusco reported having experienced physical or sexual violence from their partners. Although most studies on violence against women focus on violence that happens within the home and relationships with intimate partners (UN Women n.d.) [88], violence is not confined to these spaces: it permeates all areas of women's lives. It is true that dramatic cases of femicide have gained considerable public attention, but sadly the multidimensional nature of gender-based violence (in particular, within rural communities) is seldom discussed.

This chapter first and foremost brings to light the experiences of women human rights defenders (WHRDs), an invisible group besieged by violence and impunity (Mesoamerican Initiative of Women Human Rights Defenders 2016) [33]. Second, recognizing that research into the roles played by WHRDs during mining conflicts is an area that receives little attention (Jenkins 2014a) [54], this paper concentrates on rural and indigenous women defenders resisting resource extraction projects as a particular category of analysis. I argue here that exposing the violence and marginalization which this group of WHRDs faces in the context of their anti-mining work contributes to our understanding of how the social, political, and economic order reproduces gender violence at "the micro and macro level" (Oliver et al. 2015) [71], violence which not only heightens activists' risks but also undermines their efforts to achieve social and economic development.

The study of anti-mining activism among rural and indigenous women (Mercier and Gier 2007 [65]; Jenkins 2014a [54]; Rondon 2009 [79]; Facio 2016 [33]; Figueroa Romero et al. 2017 [32]) is a relatively new debate that has been taking place mostly among international NGOs working on the protection of human rights defenders (HRDs) and focusing on women, among them Unión Latinoamericana de Mujeres (ULAM), Latin American Mining Monitoring Programme (LAMMP), Front Line, Peace Brigades International (PBI), and the Meso-American Initiative of Women Human Rights Defenders, to name a few. Although each organization has its own approach, together they produced a new wave of literature providing evidence of how states and corporations colluded to create a violent environment which restricted activism and controlled WHRD. Thanks to national and international advocacy efforts led by the aforementioned organizations, we learned more about the crippling effect of violence on women's human rights defenders, as well as the need for countries, multilateral institutions, and corporations to develop social and protective legal frameworks responding to the specific gender needs of women defenders challenging irresponsible mining policies and operations.

This chapter focuses on Peru because for decades the main driver of Peruvian economic development has been large-scale mining operations. Not only has the country been the third-largest recipient of mining investment in Latin America since 1996, but in 2016 it became the second-largest copper and silver producer in the world (US Geological Survey 2018, pp. 53 and 151) [89]. In spite of its success, large-scale mining has not been exempt from problems which have left a legacy of environmental contamination, poverty and dependence on resource extraction. Furthermore, the mining companies' lack of attention to gender issues has resulted in a failure to deliver the promised economic opportunities for women, and the high number of social conflicts, protests, and killings of rural and indigenous activists challenging irresponsible mining policies and practices has had a detrimental impact on Peru's image as a liberal democracy (Peru Support Group 2007 [75]; Oxfam America 2009 [72]; Global Witness 2014 [42]). Regardless of its record as a successful economic performer in the developing world, one of the most significant challenges Peru currently faces is its high level of tolerance of violence against women as recognized by the Peruvian Ombudsman's Office (Defensoría del Pueblo 2015) [24]. Given the importance of mining investment and resource extraction to Peru, the use of gender-based violence by both the state and mining corporations has crucial implications for human rights, as well as activists' efforts to escape the environmental degradation associated with extractive operations.

This chapter draws on long-term research and strategic support provided by LAMMP to women human rights activists in Peru, Ecuador, Bolivia, and Venezuela, and as such it is a tribute to their commitment in bringing to the fore the illegal persecution of women defenders, as well as their lack of voice and power. It is based on observations in the field during national and regional gatherings where only women anti-mining activists participated, as well as in-depth interviews with over 100 women with whom I was personally involved for nearly two decades 1998–2016 in my capacity as Project Director for LAMMP. Prior to the start of meetings, we agreed on the questions and I secured their consent for recording and taking notes. All formal and informal interviews were carried out in Spanish and have been translated by the author. LAMMP is a London-based organization which enabled rural and indigenous Latin American anti-mining women activists to build their capacities and gain international leverage through the creation of their own national and regional networks.

To put the activism of women into a historical context, I first discuss the emergence of rural and indigenous women as activists, identifying moments of opportunities when as activists women forged fragile alliances with the indigenous movement, as well as the pervasive influence of gender power relations on the anti-mining social movement and its role in undermining and trivialising women's aspirations to become agents of social change. I then examine the experiences of three women anti-mining activists. The severity, persistence, and far-reaching impact of different forms of violence is exposed in each of the three cases selected. Notwithstanding the differences, what remains constant in each case is that during periods of mining conflict, community violence was the tool used to derail women's activism. Furthermore, the women's stories are used as a background that illustrates the pervasive nature of structural forces, how different forms of corporate, state, and community violence interact and overlap with each other, and the challenges this poses to my analysis given the paucity of research studying gender-based violence[1] in the context of mining.

Throughout the discussion, the questions that guide this paper are how gender-based violence (GBV) shaped women's participation in anti-mining activism, and more specifically how different dominant sectors used gender violence for their own benefit, in particular why violence against WHRD in rural communities is seen as "natural and normal."

5.2   RURAL AND INDIGENOUS WOMEN'S JOURNEY AS ANTI-MINING ACTIVISTS

With any attempt to explore rural and indigenous women's involvement in anti-mining activities, a significant stumbling block is the limited data about their numbers and the silence surrounding the relevance of their anti-mining activism. The emergence and recognition of rural and indigenous women as defenders of rights impacted by resource operations appears to be tightly intertwined with the latest mining boom in Peru, which World Bank researchers (Loayza and Rigolini 2016) [60] identify as starting in the 1990s when the value of the country's mining exports doubled. Social conflict and resistance to large-scale mining operations created a wave of opportunities for grass-roots social and political participation (Cornwall 2008 [20]; Bebbington et al. 2008a [5]; De Echave et al. 2009 [21]) which in turn enabled rural and indigenous women—a historically excluded group—to assert themselves as part of an emerging network of actors and community-based groups attempting to stop irresponsible mining developments.

Given their historic social and economic marginalization, it is not surprising that very little attention has been paid to rural and indigenous women's political journey into anti-mining activism. It is interesting to note that the unprecedented growth of mining investment in Peru since the 1900s has been the subject of considerable debate among academic and multilateral

---

[1]The UN Declaration on the Elimination of Violence against Women (1993) defines violence against women as "any act of gender-based violence that results in, or is likely to result in, physical, sexual or psychological harm or suffering to women, including threats of such acts, coercion or arbitrary deprivation of liberty, whether occurring in public or in private life." https://www.un.org/womenwatch/daw/vaw/v-overview.ht.

institutions. A variety of issues have been covered in a fragmented way: from the impact of mining on water (Bebbington and Williams 2008b [6]; Lust 2014 [61]), mining's contribution to reducing poverty (Loayza and Rigolini 2016) [60], governance issues, and mining impact on more than half of rural and indigenous communities in Peru (Li 2009) [59] to widespread social conflict in indigenous and peasant communities in response to mining encroachment within their communities and territories (Lust 2014 [61]; Merino Acuña 2015 [66]). However, academic interest in women's anti-mining activism and their contribution to the struggle for civil and political rights vis á vis the state and corporations has been sporadic, with a tendency to focus on indigenous women's perceived lack of agency—for example, women's conspicuous absence from negotiation and consultations.

Through my work with LAMMP, I was able to witness that women's interaction with the different actors of the anti-mining movement (traditionally male-dominated) was far from straightforward. The first issue that caught my attention was the paradox of a movement that on the one hand was advancing indigenous rights within the context of resource extraction operations, while on the other it was reluctant to take into account that the kind of social relations they were reproducing on the ground was antagonistic toward women's activists.

Kennedy Dalseg et al. (2018) [57] argue that understanding "how decision-making processes are gendered" and create inequalities is fundamental, as they influence "how development proceeds, how benefits are distributed and how negative effects are mitigated." Right from the start, an important source of gendered tension within anti-mining grass-root groups and NGOs was the internal debate about whether mining (as an economic development activity) had any impact on women, and whether women had any role in the struggle. This internal questioning played a fundamental role in undermining indigenous women's contributions and experiences, restricting them to the role of observers, as well as minimizing the importance of what they could offer to the movement. Moser's observations (2004) [69] regarding the unfair conditions which grass-roots women had to accept in order to be recognized as part of the popular communal kitchens movement in Peru are relevant to the struggle for recognition of women anti-mining activists. Moser noted that women had to "operate within a complex system of relationships … that [created] both dependency and empowerment" (2004 p. 2) [69]; a contradictory and challenging system, which on one hand, praised and empowered women as mothers, and on the other openly admitted that most leaders had to be men because family responsibilities prevented women from participating in anti-mining protests. Any attempt to challenge these rules was seen as undermining traditional values based on women's roles within the family and the principle of the complementary nature of the sexes which underpins the rich Andean cosmology.

Women activists were expected to frame their discourse in terms of the detrimental impact of mining operations on their roles as mothers. The book *Minería y Territorio en el Peru* (De Echave et al. 2009) [21] is an example of how "dependency and empowerment" worked in practice. Female and male leaders who were interviewed for the book were asked to reflect on women's contribution (but not on men's role) to the anti-mining movement, thus indirectly asking women to prove their worth as grass-roots activists. Organizations like Upside-down World (2006) [85] cited Lourdes Huanca[2] (president of FEMUCARINAP—a women's platform) as an example of how women's lack of agency to persuade the movement about the legitimacy of their concerns contributed to the establishment of unequal power relationships which privileged men and ascribed more importance to their concerns while marginalising women. Ms. Haunca openly discusses that, as part of their work, women activists had to get used to being verbally and psychological harassed by male leaders. In an interview for Via Campesina,[3] she adds that community organizations perceived women's efforts to be accepted as actors in their own right as impractical and often divisive.

Although in theory the anti-mining movement welcomed women's involvement, a constant problem was that in practice it did not see female and male activists as equal. The conviction that women's involvement would be short-lived (given their home responsibilities, among many other real or imagined constraints) cemented the perception that women were at best supporters without a specific agenda and at worst passive victims of irresponsible resource operations.

There are many examples that illustrate the gendered barriers and hierarchies built within the anti-mining movement to politically resist the advance of women activists representing local movements, while reinforcing their image as passive victims. In spite of her unique position as elected president of the "Committee of People Affected by Mining" in her town and her leading role in demanding justice at an international level, Margarita Pérez Anchirriaco was seen by the renowned "National Coordinator of Peruvian Communities Affected by Mining" CONACAMI (2009) [19] first and foremost as a woman without any organizational autonomy and second not as an activist but as a "victim of state violence."

Lack of time and resources, coupled with the constraints of participating in an anti-mining movement where men activists outnumbered women and where power was concentrated in male leaders who basically saw women as supporters and volunteers (given women's ability to provide behind-the-scenes logistical support), influenced women's decision to justify their involvement in anti-mining activism through motherhood and their responsibilities as carers of the family. Rosa Amaro,[4] the charismatic president of the "Movement for Health" of "La Oroya"—a mining town in the Peruvian Andes which during decades suffered record levels of pollution at the hands of a metallurgical complex owned by Doe Run Peru (part of the U.S. Renco Group)—framed her involvement in the movement in terms of being "closer to the children" (De Echave et al. 2009 [21]; Oxfam America 2011 [73]).

---

[2]Upsidedown world: Lourdes Huanca "the first year was really tough, because the mixed organization didn't understand and they mistreated us psychologically and verbally. They told us we were traitors and that we were betraying and dividing people. All that gave us more strength to face things, and we formed the consulting committee ..."

[3]"[En las organizaciones mixtas los hombres se han visto obligados a reconocer que] tenemos derechos a tomar decisiones ... no es solamente los machos ... Es difícil para ellos entender que tambien las mujeres saben plantear propuestas ... que tienen el cerebro para plantear propuestas ... creen que no podemos hacer debate sobre lo politico (1:53)."

[4]Amaro, Rosa: "Las mujeres estamos mas cerca de los niños, de nuestros hijos .... Esa es una de los ... motivos de nuestra lucha," p. 145.

However, women activists such as Margarita Pérez Anchirraico[5] challenged the script based on self-sacrifice with women as second-class activists. In the book *Minería y Territorio en el Perú*, Margarita does not follow a feminist or an indigenous rights discourse (Lahiri-Dutt 2012) [58], instead inspired by her own experiences of the mining conflict she proposes a development path that rejects mining and is to be based on "agricultural and livestock farming." For her, "covering the land with concrete is not development." She also challenged the traditional perception of women lacking "the necessary capacity and that [activism] is not for us. They always expect us to be below them, that is wrong[6]" (De Echave et al. 2009, p. 328) [21].

Exclusion from important meetings (agreed through informal communication networks to which women had no access) and being subjected to unwanted sexual attention[7] were also identified as obstacles women had to accept as part of their work. The findings of Freidenberg and Osornio Guerrero (2017 p. 276) [39] in their study of the political participation of women in Mexico provide evidence that women who attempt to compete with men for spaces traditionally controlled by them are subject to "patriarchal barriers, misogyny, verbal, and sexual violence as well as exclusion."

To summarize the discussion, I put forward the argument that social, economic and political constraints which male-led groups erected around women had consequences for their anti-mining work that went beyond being belittled and verbally and sexually harassed. First, such limitations shaped the ways in which women could resist (Jenkins and Rondon 2015) [56]. Second, unequal social hierarchies that discriminated against women were used by companies as an excuse to discharge their responsibilities toward women, exclude them from meetings and reject women's claims that mining operations had a disproportionate impact on them. Finally, using the argument that it was important to avoid the impression that the movement was divided, women were discouraged from participating in the highly competitive world of international conferences and from building direct contacts with potential funders. This maneuvre had a knock-on effect on women's activism. Their isolation contributed not only to starving local women's groups of funds but fundamentally "… exacerbated power and resource imbalance among activist organizations …" (Alvarez 2000, p. 23) [2].

It is my view that through their own short-sightedness the anti-mining movement failed to see that by anchoring their discourse in gender issues, women were widening the scope of the movement, helping to draw attention to fundamental weakness of the resource-extraction model. On the one hand, women denounced that mining operations had transformed them into the poorest among the poor, thus challenging state and corporate claims that mining was the

---

[5]Pérez Anchirraico, M. "Pero no deberían afectar a la zona agrícola, ni a la ganadera. Pero afectan y eso no trae desarrollo, para mí desarrollo es potenciar la ganadería y la agricultura, mas no estar sembrando cemento," p. 328. Minería y Territorio en el Perú.

[6]Pérez Anchirraico, M. "…pero para los hombres, las mujeres no tienen la capacidad y no están para estas cosas. Siempre nosotras tenemos que estar a un nivel más bajo que ellos, están equivocados," p. 328, Minería y Territorio en el Perú.

[7]Huanca, L. "Las mujeres ya no son acosadas sexualmente como cuando se inició el primer congreso de la CLOC [Coordinadora Latinoamericana de Organizaciones del Campo] …Los hombres a veces dicen en broma hay que darles espacios para que no nos fastidien" (Interview/video for Causa Campesina 2:59) https://www.youtube.com/watch?v=7qp80uTuKZQ.

right engine for the sustainable development of Peru's most vulnerable communities. On the other hand, women made it evident that as a result of lacking a gender perspective, the industry had failed to deliver development opportunities for them.

In spite of what appeared to be insurmountable obstacles, women activists began to create their own spaces and networks, among them ULAM (redulam.org), which helped activists to mitigate their lack of agency and access international solidarity (Rousseau 2015) [81] Over the years, the women's anti-mining movement in Peru has had losses and victories. Two important achievements stand out: first it helped women to forge an independent identity as human rights activists; and second, through successfully claiming a space for rural and indigenous women they were able to generate a more nuanced understanding of the gendered impact of mining activities on human rights. For example, when the lonely struggle of Máxima Acuña against Newmont (Conga Mine) in Cajamarca reached the front pages of newspapers across the world, thanks to the support of ULAM and other women's groups her activism was recognized and portrayed by top international NGOs as work in defense of fundamental human rights. Maxima's work reflected well on the activism of rural and indigenous women in Peru (Celendin Libre 2014) [16] albeit short-lived, as to this day little is known about the work of many invisible women fighting on their own against mining projects operating within their communities.

5.3   FACING VIOLENCE HEAD-ON: THE EXPERIENCES OF THREE SELECTED ACTIVISTS

The analysis of women's involvement in anti-mining activism presents many challenges. It is not just its complex nature: the invisibility of their struggle means that numbers alone cannot be used to reveal the extent of their activism or the depth of their suffering. Farmer (2004) [34] noted the statistical invisibility of people who during their lifetime had experienced nothing but poverty and brutality, and championed the use of testimonies as the best way to gain insight into the long chain of violence, injustice and maltreatment which characterized their lives (a context which Galtung 1969 [40] and Gupta 2012 [46] defined as structural violence). Following Farmer's (2004) [34] preference for testimonies, I will use the stories of three women activists I worked with to explore the richness of their experiences, their views about their struggle as activists as well as their understanding of the mining agenda of the state.

Margarita Pérez Anchirraico and Máxima Acuña are two well-known Peruvian activists living in rural communities torn by social tensions and conflict as a result of mining activities. Although they live in opposite geographical areas of the country, the two women have a Quechua-speaking background and their daily routine centers around raising a family, farming, taking care of their animals as well as the occasional visit to the market to sell home-made products. Margarita was born in San Mateo de Huanchor—a small village 95 km to the north-west of Lima—where she lived with her children, mother, and brothers.

Margarita is a strong woman used to taking decisions, so when she found out that Peruvian "Minera Lisandro Proaño" owner of the underground mine and mill complex known as "Coricancha" (gold, silver, lead, zinc, and copper) located in the "Tamboraque" area (5 km south-east of Margarita's town) had illegally dumped 400,000 tons of toxic tailings from their smelter in the woods close to her house she decided to do something about it. "I was the one who talked the most, so they elected me" as president of the "Comité de Defensa del Medio Ambiente y Desarrollo Sostenible." As part of a campaign to get the mining company to remove the tailings, she visited other villages and started a network for activists who marched with her to the city. Most importantly, she relentlessly lobbied members of the government and local authorities until the Environmental Health Department of the Ministry of Health[8] carried out a health assessment in 2002, which confirmed that 48 people in San Mateo showed signs of metal contamination (lead, zinc, and arsenic), with 86% of children examined showing signs that their mental capacities had been affected (ejatlas.org). With that information to hand, Margarita approached national and international NGOs until eventually with the support of CONACAMI and the U.S. Center for International Environmental Law (CIEL) she submitted a complaint before the Inter American Commission for Human Rights (IACHR). Margarita's complaint was upheld by the IACHR[9] which instructed the Peruvian government to remove the tailings, repair the damaged land, and provide medical support to those affected (CIEL 2004) [18]. Margarita did not take time off to celebrate her triumph. Instead, she started criminal proceedings against the general manager of "Minera Lisandro Proaño," demanding that the company provide compensation and medical assistance to the community, because toxic fumes from the sludge field had seeped into the soil and were present in the air and making people ill, children in particular.

In the meantime, the Peruvian "Minera Lizandro Proaño" as a result of (allegedly) low metal prices as well as the environmental and health crisis that it had created in Mayoc, was forced into bankruptcy. As a major creditor, the bank Wiese Sudameris Leasing S.A. became the proprietor of the Coricancha mine, but was unsuccessful in its attempts to restart work, given environmental concerns. In 2006, the mine was acquired by the Canadian company Gold Hawk Inc. and managed by "Minera San Juan (Peru) S.A." a wholly owned subsidiary of Gold Hawk. According to Margarita, in spite of an aggressive campaign that aimed to get the support of the Mayoc community, Minera San Juan seemed unable to start production, given the heightened concerns about environmental contamination and the IACHR complaint. Subsequently in 2009, Belgian Nyrstar bought and managed the mine until June 2017, when it became the property of Canadian Great Panther Silver LTS.

During interviews, Margarita stated that she began experiencing verbal abuse, threats and persecution in 2006 when "the company Minera San Juan tempted men into abandoning their opposition through offers of well-paid jobs, training and hand-outs to the village." Soon afterward, the community turned against Margarita, requesting that she withdraw the health and environmental accusations she had made before the IACHR. Margarita also started receiving warning messages. Sometimes it was just short, written messages slipped under her front door

---

[8]DIGESA in Spanish.

[9]Perez Anchiraico v. Peru, Inter-Am. C.H.R., OEA/Ser.L/V/II.127, doc. 4 rev.1, Ch. III, para. 43, 2006.

advising her to take care because she was in great danger; at other times her electricity was cut off, and on one occasion a large hole was dug outside her home and a black cross with the words "RIP Margarita" put in the center of the hole. Although Margarita continued with her activism and reported to the police every threat and message, she felt that "their indifference was extraordinary."

Finally, Margarita was forced to abandon her activism after a mob of about 30 men from her village attempted to lynch her and throw her into the turbulent waters of the nearby Rimac River. The IACHR was alerted to this attack and requested that the government take measures to protect her life. Unfortunately, police protection was short-lived, because Margarita could not afford to cover the costs of transport and food of the officer assigned to her. During an interview in 2016, Margarita spoke with regret about the years lost in activism, confiding that memories of the way she was treated by members of her community still haunt her. No longer fired by the passion and energy of her early days as an activist, Margarita looked back with sadness, given the emotional scars left by the abuse and persecution. "I was singled out as the person responsible for depriving men of work. I was lucky, they could have killed me …. After the attempted lynching I was isolated as though I were a leper …nobody talked to me. Public services were denied to me. On reflection, I can see that my activism was only temporarily accepted. The men didn't want a woman telling them what to do. I paid a very high personal price for my work.[10]" Margarita continues to live in her community.

Through her activism, Máxima Acuña achieved fame and recognition (The Guardian 2016) [45], with more than 50 entries in Google written in English, French, German, and Russian. A traditional weaver and subsistence farmer who sold cheese and hand-made blankets, I met Máxima Acuña during ULAM's AGM celebrated in Lima in 2014. During this event, Máxima disclosed her personal circumstances and her struggle to be recognized by Minera Yanacocha S.R.L. as the rightful owner of the land she occupied in Tragadero Grande (district of Sorochuco), a small village in the highlands of Cajamarca. Maxima's misfortune started when she found out that the land she had previously bought and converted into her family home was included as part of the expansion of Yanacocha—South America's largest gold mine, and a joint venture between U.S Newmont Mining Corporation, Peruvian Minas Buenaventuras and Japan's Sumitomo Corporation. Like Margarita, Máxima fought for her rights for many years, and her international campaign took her to Europe and the U.S. Between 2014 and 2016, I carried out several interviews with Maxima at her home in Sorocucho—a tiny village high in the mountains of Cajamarca—where I was able to witness how her property, home, and movements were under constant surveillance by the company. I also carried out interviews with all members

---

[10]Pérez Anchirraiaco, M. (Cuzco, June 2016): "Me querian matar … Me declararon la persona responsible de quitarle a los hombres su trabajo …. Tuve mucha suerte y si no es por la ayuda de mi sobrino me hubieran tirado al río …. Despues del intento de lincharme me converti en una leprosa para el pueblo … nadie me hablaba hasta el dentista se negó a prestarme servicios. He llegado a la conclusión que me aceptaron solo por unos meses mientras podian ver los deshechos en el bosque …. Aunque no sabian que hacer los hombres nunca aceptaron que una mujer opinara … Me duele por mis hijos … porque los descuidé. Hoy estoy sola y siento que pagué un alto precio y logré muy poco."

of her family, and shared accommodation with her and her daughter during a two-week tour of Europe organized by LAMMP.

Her hopeless beginning as an insignificant Peruvian indigenous woman fighting to keep a piece of land which Newmont claimed it had previously acquired, earned Máxima the sympathy of foreign governments, international NGOs, and around 5,000 followers on social media. Like Margarita, Máxima's struggle for what she referred to as "crumbs" went on for years. Within her community she was harassed, publicly humiliated, and intimidated. The agony of her suffering seemed to have no end. "They [the police] beat my son with their machine guns, they attacked my husband, they beat my daughter-in-law and my daughter. When I saw that my daughter had passed out, I went to help and three policemen grabbed me by each arm. I lost consciousness when they were hitting me with their stick" (Acuña 2012) [1]. The story of Máxima Acuña is characterized by criminalization (judicial harassment) as a form of corporate and state violence. In spite of her meager resources, Máxima was frequently forced to travel to neighboring towns where she had to attend court hearings that often were adjourned at very short notice, and where she had to defend herself from claims that she had participated in acts of violence against the mine. In 2011, Newmont announced that it had suspended the Conga Mine development indefinitely. In 2016, Maxima was awarded the Goldman Environmental Prize, but it was not until 2017 that the Supreme Court resolved the land dispute between Maxima Acuña and Newmont in favor of her family. Maxima still lives in her home in Sorocucho, Cajamarca.

The last woman informing this paper is Rut Fernandez, a young mother of two living in the village "Alto Coñicorgue," Huasmin District an hour and half from the city of Cajamarca, capital of the region of Cajamarca in the northern Andes of Peru. As a supporter of the platform "Women for Life Association" (Asociación Mujeres por la Vida) based in Cajamarca, Rut is part of a first generation of rural women well-versed in the detrimental impact of mining on water quality and availability. On the other hand, as head of a small family which she supports through her subsistence work, she felt free to make her own choices, including taking part in communities' mobilizations to protect water sources targeted by the Conga mine: an expansion of the Yanacocha mine that fuelled numerous protests, confrontations, and pitted grass-roots campesino groups and the regional government of Cajamarca against the central government. There is evidence that these protests were met with excessive force as well as widespread arbitrary detentions and beating of unarmed individuals (Frente de Defensa Ambiental de Cajamarca 2011 [38]; Diario NCO 2011 [25]).

Prior to Rut's first direct experience of the brutality of four neighbors, she recalled (in a private interview, March 2018) that her four attackers (who had previously worked for the Yanacocha mine, were unemployed at the time, and favored the expansion of the Conga mine) had sent her frequent and threatening phone messages warning her that protesting against Conga was off-limits to her. Although Rut felt under surveillance during protests, as her neighbors were never too far away from her she did not think her neighbors would dare to carry out their threats. Like many other women, Rut's understanding of GBV is narrowly defined as the physical and intimate violence that occurs within the limits of the home, and she failed to perceive the threats, stalking, as well as attempts to control her and dictate what she could do in her private time as another form of violence (Galtung 1969) [40]. During interviews, Rut reported that it wasn't unusual for men in her tiny village to challenge her activism and question her apparent freedom.

During a phone interview in June 2017, Rut explained that the makeshift house in which she lived was the property of her family, and that she shared it with her sister Fanny Fernandez. Regarding the attack, she recalled that about eight in the morning of the December 15, 2015 together with her sister she was cooking breakfast, when four neighbors abruptly walked into their house, enquiring about their brother. Rut noticed that one of the men was carrying a machete, but thought he was on his way to work. When she replied that her brother was not in, one of the men grabbed her by her ponytail and hit her with a stone. While she was on the floor, the men kicked her and her sister violently. Fanny was repeatedly stabbed in her face and above the waist area. They were saved from what they felt was imminent death by Felipe Fernandez (a male cousin who had arrived the night before), who came out of his room in their support. Felipe was attacked with the machete and suffered injuries, but the men ran away. Four days later while Rut, her sister Fanny, and their cousin Felipe were in hospital about five o'clock on the afternoon of the December 19th her neighbors returned to the house and burned it to the ground. According to Rut, the intention behind the destruction of her home was to force her to move out of the village, which in turn would enable her attackers to grab her land, move in, and eventually sell it to the mining company.

Once out of hospital, Rut went to the police to request that the incident be investigated. Notwithstanding the fact that Rut identified her assailants and showed photos with her injuries, the police refused not only to file the incident but also to take any action against them. Her attempt was dismissed on the grounds that she should have reported the incident earlier, so forensic evidence could be obtained.

The police decision meant that if Rut wanted justice, her only option was to take legal action and sue the perpetrators directly, which she did with the financial and emotional support of her group.[11] In 2018, after finding the perpetrators guilty of physical assault that caused "superficial" injuries[12] to Rut, Fanny, and Felipe, the perpetrators were sentenced to three years and six months in prison. However when passing sentence[13] the judge felt obliged to consider that the brothers had pleaded guilty and did not have criminal records and in view of this, their sentence was suspended. In addition, the four brothers were sentenced to pay in compensation 7,000 Peruvian Soles (approximately 3,382 U.S. dollars) which took into account physical injuries as well as destruction of property. Furthermore, taking into account that Rut's neighbors

---

[11]Expediente 00184-2016-0-0603-JR-PE01. Celendin September 27, 2017. Judge Edwin Sergio Chacon-Nuñez.

[12]Article 122 of the Criminal Code establishes that injuries caused to a person that require between 10 and 30 days of hospital care should be considered superficial ["leves"].

[13]Sentence No. 12-2018 Celendin, January 24, 2018.

lacked financial resources to pay the fine in full, it was agreed that they could pay it in two equal instalments which would have to be made one in April and the other in July 2018.

Beyond the privacy of her family, Rut's attack was not discussed, perpetuating the underlying structural nature of GBV. Rut resented that the violence she experienced slipped past unnoticed by her community, as well as by the local police. To Rut, the decision of the criminal justice system to free her attackers confirmed that violence against women was "normal." Lack of accountability is for Boesten (2012) [10] a key factor that fuels violence against women. After reviewing statistics for her study of violence against women in Peru, she concluded that in Peru "few reported cases are processed judicially and even fewer lead to conviction" (2012 p. 362). Freidenberg and Osornio Guerrero (2017) [39] eloquently define impunity as a complex web of practices based on "silence, omission and negligence in collusion with the authorities responsible for preventing and punishing violence."

Although Rut considered appealing against the sentence, she decided against it for fear that her neighbors might retaliate. However, her group used her experiences for capacity-building training. Their conclusion was that their aspirations to lead social change exist within a patriarchal system that denies them the possibility of dreaming about rights, but nevertheless appears to be more tolerant than in the past. That women operate on shifting ground is recognized by Horowitz (2017) [51], who points out that far from being static, rural, and indigenous women's social positions are in fact complex, dynamic, and evolving. Rut left her village and now lives with her sister in another area.

5.4   FACTORS THAT CONTRIBUTE TO GENDER-BASED VIOLENCE IN THE CONTEXT OF RESOURCE EXTRACTION OPERATIONS

5.4.1   PATRIARCHAL FORCES

When reviewing studies about GBV, the starting point of a great number of them is that it is underpinned by power and privileges for men (Flood and Pease 2006) [36]. Heise (1998) [50] considers that this context of male entitlement serves as the base for gender norms, in turn responsible for creating a patriarchal framework and a punitive "social order" which coerces women into compliance (Jakobsen 2016, p. 3) [53]. In the case of rural and indigenous women, Figueroa Romero et al. (2017) [32] adds that because these factors overlap and reinforce each other, they create multiple layers of exclusion and discrimination.

The activists I worked with had no doubt in their minds that the multiple forms of marginalization and violence they experienced growing up and later as activists were an expression of the "machismo" men "picked up" within their traditional families. They understood "machismo" as the accepted social order which made men superior to women (Castro and Riquer 2003) [15]. Describing the power and control men had over their lives was accompanied by expressions such as "that's the way things are." The inevitability of their situation always brought to my mind the Star Trek TV series that popularized the expression "resistance is futile."

Given the constraints arising from women's subordinate position, activism was seen as a possibility only for a handful of exceptional women. As discussed, strict norms confined them to the private space of the family. Traveling alone to participate in meetings held outside their communities created problems at home. An additional factor limiting women's activism was their low educational levels and lack of identity papers (common among indigenous women). Gender restrictions coupled with police harassment of women identified as activists transformed the act of travelling alone into a highly personal risk. But what women reported as demoralizing was to find out that the organizers of the event had allocated to their presentation just a few minutes, and usually at the end of the meeting. To them this practice meant that for the mining movement their participation in public meetings was symbolic.

Although a generalization, the essence of living in a "macho society" was summarized as working hard (laziness in a woman was not tolerated) and accepting that financially they depended on their husbands who were responsible for both the small as well as the big decisions—for example, number of children. In general, women were expected to obey and respect all male members of the family, and behave in a way which did not compromise or challenge their authority. Activists with family members in the police or the army reported conflict and violence within their families because they were expected not to participate in protests or speak against mining projects.

I found the work of Heise (1998) [50] on gender norms that create a patriarchal structure dictating how women should live their lives, Jakobsen's (2016) [53] arguments regarding social order and Figueroa Romero's (2017) [32] analysis of how patriarchal normative frameworks reinforce each other relevant to my understanding of women's lack of choice and the severe impact of violence on their activism. However, alone these theoretical perspectives did not explain the full spectrum of violence encountered by women activists.

Drawing from the work put forward by many other theorists, in the next sections I endeavour to produce a theoretical framework—grounded in women's testimonies—(Menjivar 2008) [64] that helps to unravel the violence activists encountered in the context of their anti-mining work.

5.4.2   GEOGRAPHIC LOCATION, GBV, AND MINING

De Keseredy (2019) [22] considers geographic location as a key predictor of rural and indigenous women's higher vulnerability to violence. "Rural women are at higher risk of experiencing [violence] than are those in more populated places" (De Keseredy 2019, p. 316) [22]. International agencies such as WHO (2005) [92], Pan American Health Organization (2012) [74], and national entities such as the Ministry for Women and Minorities (MIMP 2009) and the National Institute for Statistics and Information (INEI 2018) [52] provide evidence that indeed for rural women in Peru, violence is a common occurrence. Through their work in Australia,

Kerry Carrington et al. (2010) [14] make the connection between geographic location and the "hyper masculinity" of the men working in mining operations. The authors raise concerns about "the high rate of violence among men …. working and living in communities experiencing rapid expansion due to global resources boom" (2010, p. 1) [14]. Carrington et al. explain the "male-ness of violence" (2010, p. 1) usually associated with mining towns in the excessive valuation of male culture and male values traditionally associated with the image of mining as an activity for "real men," a stereotype that contributes to "sexist male subculture dynamics," responsible for rewarding maleness and privileges for men.

Several aspects of the framework put forward by Carrington et al. are useful for studying the extent to which mining extraction has transformed rural communities in Peru and exacerbated violence against anti-mining activists. The starting point is that rural areas which have experienced a dramatic increase in mining operations (Bury 2005) [13] are mostly located in the Peruvian Andes, a region with a high proportion of people from an indigenous background (INEI 2018) [52]—a sector traditionally neglected by the state. Even though the expansion of the resources sector in Peru has contributed to the economic growth and social transformation of rural and indigenous communities, nevertheless very little has changed in term of public service provision and the quality of lives of indigenous people. Furthermore, Peru has failed to overcome inequalities within groups, with rural and indigenous women considered as the poorest social category in the country. A study commissioned by the ILO found that rural communities traditionally engaged in agricultural and livestock activities continue to have high levels of poverty. "While the poverty level at the national level was 22.7% in 2014, in that same year, among peoples whose mother tongue was Quechua or Aymara, it was 34.1% and, among those with other indigenous languages, the rate was 64.7%" (Del Aguila 2016, p. 3) [23].

In my experience, most women anti-mining activists I worked with were of indigenous extraction and lived in rural communities in close proximity to mining operations. The traditional, rural villages in which they lived were tucked away in such remote and geographically isolated areas that even with private transport I had to put a full day aside whenever I visited them. A picturesque landscape made up of makeshift houses scattered on the mountains were the only sight visible to visitors. What distinguished these communities from others is that although families are often related to each other, they depend on no one for their survival but their subsistence work. In these communities, absence of the state means lack of essential public services—running water, phone, internet, transport, schools and health centers. Isolation coupled with remote location and the absence of technological advances meant that activists lacked the tools to publicize their resistance or report any violent attack against them. Furthermore, mining companies' practices of blocking or diverting access to houses adjacent to their operations contributed to the invisibility of these geographical spaces. In turn, these conditions made it difficult to support them when in danger. Paradoxically, their low number meant that activists were easily identified, which in the event of wanting to file a complaint against a perpetrator of abuse made them vulnerable.

5.4.3   ETHNIC IDENTITY AS A RISK INDICATOR OF GBV

A defining characteristic of rural communities in Peru is their ethnic background; of the indigenous people living in the Peruvian Andean over 32% live in rural Andean areas (INEI 2018) [52]. In 2018, a study commissioned by the International Development Bank (BID 2018) [8] considered women's ethnic background as a "high risk" predictor of violence. In particular, the report shows that in Peru violence is high in families "in linguistic transition" (2018 p. 7), that is families where Spanish is currently used but growing up the mother tongue of the family was an indigenous language. Although the majority of women anti-mining activists were fluent in Spanish, a significant number remembered being raised in families where only an indigenous language was spoken. For them, that their children learn Spanish and attend schools where Spanish is taught has become a priority.

Authors such as Boesten (2012) [10] and Drinot (2011 [26]; 2014a Note 10, 11 [27]) hold the view that as a nation, Peru struggles with its indigenous identity; this in spite of Peru being "in terms of the overall size of [its] indigenous population the second country in Latin America with the largest proportion of indigenous people" (UN Expert Group 2018, p. 3) [86]. For Boesten, ethnic background is essential to understanding gender-based violence in Peru, a country which she defines as "highly biased according to race and class" (2012 p. 17). I believe the extensive testimonies provided by the activists—in their capacities as rural and indigenous women—offer support to Boesten's theory that violence against women has to be understood against this backdrop of historical gender and race discrimination that places rural and indigenous women at the bottom of the Peruvian social structure. I would add that in the context of their anti-mining work, activists' vulnerability to GBV is exacerbated by their determination, first to reject the accepted image of rural and indigenous women as "passive victims" of resource extraction projects; and second their resoluteness to use their experiences to curtail the advancement of an extractive model of economic development that excludes them. To me, rural and indigenous women's determination to fight on the ground against irresponsible mining operations earned them the wrath of successive governments in Peru who vilified them. Paulo Drinot (2011 [26]; 2014b [27]) puts forward the argument that the identification of certain sectors of the country as "obstacles to national progress" and "enemies of the country" obscures the fact that the real issue at stake is racism directed against "a large proportion of the population" (2014b p. 12) [27]. Former president Alan García Pérez 1985–1990 who in his second term in office 2006–2011 pushed for an economic development model based on trade and natural resource exploitation is considered by Drinot (2011) [26] as the best exponent of "a war against indigenous people" who opposed his efforts to open up the Amazon region for logging, mining, oil and gas exploitation through presidential decrees in 2008. In a series of articles published in national newspapers, president Garcia considered that "natives are blocking Peru's road to economic success in the 21st century" (Marti 2012, p. 32) [63]. Through an analysis of Garcia's TV speeches, Escobedo (2016) [31] argues that "...the most ordinary discourses in Peruvian politics establish and normalize a structure in which racism can flourish, persist, and become a central element in, for example, social conflict in relation to oil or mining concessions" (2016 p. 260).

5.4.4 IMPACT OF MINING OPERATIONS ON RURAL POVERTY

Against a backdrop of historical poverty, it is difficult for a rural family to survive entirely on farming alone. As jobs in rural areas are scarce, an obvious solution is for men to gain employment as laborers on wealthier farms, and for women to seek employment in domestic service within private households. In this context, the importance of the mine for local jobs is significant, not only because it is one of the very few sources of employment but fundamentally because it offers workers income and benefits that are not found anywhere else. It is also a source of division and resentment within the community, in particular, because very few unskilled men will be chosen and rewarded with work at the mine. In a review of employment in the mining industry using data from MINEM, Sanborn and Chonn (2015) [82] reported that MINEM estimated that around 53% of employees were members of local communities, 46% from other regions, and just 0.26% foreigners. They also noted that a higher portion of those hired are employed by contractors rather than by the mining companies themselves. Contractors tended to have lower salaries than mining company employees, and usually experienced a greater vulnerability in their rights. Sanborn and Chorn also quote MINEM in order to estimate that by 2013 subcontractors represented 67.4% of total direct employment in mining. It is from the perspective of the higher earning power and social standing associated with the competition for mining jobs that are scarce that I see how it can become a source of GBV. Margarita and Rut's testimonies provide evidence that women's anti-mining activism was perceived by local men looking for work with a mining project as a particular affront to them, an outrage that was quickly rectified through the use of violence. In her paper on the mining industry in Peru, Armstrong et al. (2014) [3] provide evidence of the importance of the mine as an alternative source of work for men, and how mine engineers with offers of working for the mine manage to entice men to help the corporation secure the support of those who don't want to sell their land or are antagonistic to mining operations.

5.4.5 MINING WORKERS' IDENTITY AND VIOLENCE AGAINST ANTI-MINING ACTIVISTS

Carrington et al. (2010) [14] argue that the traditional image of mining as an activity for "real men" appeals to rural men who usually take pride in working hard. In remote, rural communities of the Peruvian highlands working for the mine is part of a privileged male world that rewards hard-working men with respectability, competitive salaries and the possibility of acquiring new skills—for example, learning to drive big machines or to use heavy equipment which are markers of masculinity. Like in any other enterprise, men brought together to work for a mine develop loyalty, connectedness to the project, a sense of belonging and take pride in being part of a clearly-identified group with a shared purpose. Shared values around mining and the culture of the company allows men to form strong bonds and reinforces typical "macho" male behaviors common in rural areas. In this context of heightened masculinity coupled with privileges that set them apart, it is not unusual for rural mine workers to invest heavily in their work and by extension develop a collective interest in the mine. In a study of male workers' violent behavior in Guatemala, I examined how these factors interact and how under peer pressure rural men from the community organize groups that patrol the community and feel entitled to attack gatherings where the community is discussing the impact of the mining company (Rondon 2011) [80]. Margarita Perez's testimony of being violently attacked by a mob of workers who feared that with her activism she was endangering their jobs is an example of that same behavior in Peru.

It is from this perspective that I argue that the violent behavior of mine workers and the role played by mining dynamics at the community level has been underestimated. Through their resistance to resource extraction projects, women activists living in rural communities are at a higher risk of violence not only as a result of the interplay of gender discrimination, patriarchal forces, geographic and social isolation and their indigenous background, but because the operational activities of a mining project lead to a redefinition and reaffirmation of violence as both normal and an essential identity marker of real men. When men's privileged status is threatened (especially by women), "violence can be a way of reinforcing boundaries, exercising power, asserting male honor, and re-establishing social status with other groups of men" Carrington (2010 p. 10) [14]. The testimonies of the activists discussed in this chapter show that rural men working for the mine expected women to become subordinate to their needs as workers. Failure to do so was seen as compromising their domestic authority and diminishing their standing—among peers and the community—as real men. Through this logic, women became legitimate targets that have to be punished, and through a climate of fear and uncertainty, made an example of (Rut's testimony).

5.4.6 STATE AND CORPORATE VIOLENCE

Although mining-related mobilization of excluded sectors—rural and indigenous people in particular—did not achieve a shift in the state's priorities, there is recognition that in Peru these protests contributed to a repositioning of policies and changes to the way government and corporations engaged with sectors traditionally defined as marginal (Rénique 2009) [78]. Sanborn and Hurtado (2017) [83] argue that given Peru's dependence on natural resource extraction, governmental changes aligning the state and its export sectors with global trends was inevitable. In this context of economic cooperation, facilitating mining operations, mining investment, and the continuous expansion of the sector became the goal of the state (Li 2009) [59]. The future development of Peru as an advanced, modern nation was cemented by a political, populist, and nationalistic ideology that directed its firepower against those who opposed extractivism (Vom Hau and Biffi 2014) [90].

In this section I argue that at the time the activists were opposing mining developments, the Peruvian government and the mining industry were under intense national and international criticism for their tendency to respond with excessive force to protests. The government's eagerness to impose resource extraction operations as a "win-win" development model for Peru coupled with its lack of consideration of the right for self-determination of indigenous people (UN Special Rapporteur for Indigenous Peoples 2013) [87] earned the country a poor reputation as a democratic country (Fougere and Bond 2018 [37]; Garita Vílchez 2013 [41]). On the other hand, as part of global efforts to maintain the legitimacy of mining operations, Peruvian corporations felt the urge to engage with communities and become a "good corporate citizen." With time, the need became evident to these two key players to join efforts and work together in what Brock and Dunlap (2017 p. 34) [11] define as "corporate counterinsurgency," that is an integrated approach capable of weakening the impact of mass protests but within the scope of the law. Although the favored government strategy of deploying military forces (militarization) to communities engaged in opposition to mining operations has not been abandoned, criminalization of activists, as a soft strategy, has become the preferred option to nullify unwanted opposition. Today, criminalization is so widely used that it has become "normal," while its dependency on "the rule of the law" has made invisible its darker, repressive, violent nature. By repeatedly using judicial harassment and persecution, the state and mining companies have not only forced (male and female) activists to fear the consequences of their activism but more importantly have made criminalization a normal occurrence, something that almost inevitably will happen to activists. Through this process of "normalization," a new form of state violence becomes invisible, a process that Boesten (2010) [9] conceptualizes as "normative violence …violence that is socially not understood as violence because of its normalization; it is tolerated and normalized because it is perpetrated in response to social transgressions" (2010 p. 5). As a strategy, criminalization has united both the Peruvian state and corporations, providing them with a less confrontational and more legitimate approach against popular narratives challenging extractivism.

Notwithstanding its widespread use, criminalization alone did not yield the expected outcome of eliminating the grinding unrest of mass protests and rallies threatening the smooth operations of mining corporations. The need to put new strategies in place "to stabilize and mitigate conflict in areas of interest" became evident (Brock and Dunlap 2018) [11]. Through changes in legislation, the state-enabled large scale-mining corporations to directly engage police forces—including the special forces DINOES—in the defense and protection of corporate interests and operations (Armstrong et al. 2014) [3]. The significance of this new alliance cannot be overestimated as historically, repressive violence through the army and the police have been the exclusive monopoly of the Peruvian state, and Fujimori's reliance on the use of state resources and violence to annihilate the Shining Path guerilla provides evidence of this legacy (Burt 2006) [12]. The importance of these legal developments—kept under wraps for some time—cannot be overestimated as they show that different democratically elected governments felt compelled to share and facilitate corporate access to state violence, in order to ensure the advancement of the extractive model (Grufides 2013) [43]. Inevitably, these legal developments draw attention to the decisive and influential role of mineral extraction operations in Peruvian governance and its contribution to the weakening of the judicial system in Peru. Drinot (2014a) [27] argues that the judiciary system in Peru is perceived as corrupt by the majority of the population.

Newmont was among the first mining corporations to benefit from signing legal agreements with the police forces, co-operation which increased the impact of criminalization. Borrowing from the conceptualization that Brock and Dunlap (2018 p. 34) [11] make of "corporate counterinsurgency" as a complex strategy used by corporations to win over communities, in the next paragraphs I examine the complex master plan that Newmont deployed against Maxima Acuña—whose position against the development of Newmont's "Conga" mine had made her into a leading figure of the anti-mining movement in Peru. From this perspective of "counterinsurgency," I argue that Newmont's corporate strategy shows perfect synchrony among different players, smooth operations between state structures and the company's own security personnel, in general efficiently coordinated at the local and national level with the sole objective of suppressing anti-mining opposition in Cajamarca. Maxima Acuña was an unknown farmer whose personal circumstances achieved notoriety after a video of a failed eviction of the family from their farm went viral. The video (Acuña 2012) [1] showed security personnel and a platoon of around fifty officers from DINOES (police special forces) brutally assaulting Maxima, her husband, and their two daughters as part of an eviction requested by Yanacocha. The harrowing video of a family running and screaming for help while a representative from the prosecutor's office watched passively and undaunted at what was happening shook the nation. Although Maxima reported the incident to Newmont and the local police, and also managed to persuade a lawyer from the prosecutor's office to see for himself how the police had destroyed her property, to this day none of her complaints have been upheld (Red ULAM 2014b) [77], thus confirming that the violations of HR reported by Maxima were not perceived as important by the authorities. After police forces failed in their attempts to evict the family, Yanacocha used the judicial system to intimidate Maxima with threats of a long prison sentence unless she voluntarily left the land (Grufides 2015) [44]. Heavily criminalized for her activism, together with other anti-mining leaders, for several years Máxima was forced to periodically travel from Cajamarca (the nearest city) to Chiclayo (Lambayeque), where their legal processes had been transferred. Traveling with her family added another layer of punishment and suffering to Maxima. "In addition to being forced to travel for at least four hours by bus, often when we get to Chiclayo we are told that the hearing has been canceled or moved to the following day which means that we have to pay not only for transport but also for a place to stay the night." For Máxima and her family these changes "were deliberate, and reinforced the feeling that we are powerless." The "transfer of jurisdiction and the systematic archiving of activists' claims" are considered by Guzman Solano (2016) [47] as part of a strategy of "legal marginalization" deliberately created by the Peruvian state not to enforce the law but to punish people for their anti-mining views.

Brock and Dunlap (2018) [11] state that a "corporate counterinsurgency" approach to environmental activism also involves "infiltrating the social fabric" of communities as another way of invisibilising "harder" repressive techniques. During interviews, Maxima often spoke of tensions within the family as well as in the wider community as a direct result of state and corporate persecution. Although difficult to prove, Máxima provided examples of community pressure which she linked to men working for the mine, e.g., anonymous messages with death threats, conductors of local buses who she alleged refused to let her board for fear they might lose the business of people linked to the mine, attempts to kill the family dog, as well as destruction of her bike. Finally, the legitimacy of her struggle was questioned by anonymous people who spread insidious rumors that she was receiving funds from international NGOs, or that her interest in remaining in her home was to secure a substantial hand-out from the company. In addition, there was evidence that as part of "a culture of fear" the company was using its own security personnel to destroy parts of her property and crops, to maintain Maxima and her family under constant surveillance by security cameras, and to restrict her movements. In 2017, after more than five years involved in a long legal battle, the Supreme Court of Peru acquitted Máxima Acuña of the charges of illegally occupying her land.

Let us now turn attention to the way in which state institutions condone violence against women. By the simple act of doing nothing to prevent it, the state is complicit in all forms of violence against women. Criminalization of activists raises questions regarding the Peruvian government's expressed commitment to the elimination of gender violence as an obstacle to equality. Does a state keen to attract mining investment benefit from gender norms that suffocates women's anti-mining activism? The fact that violence against Maxima, Rut, and Margarita wasn't questioned within their communities shows how useful are gender norms when it is necessary to limit and control the power of women activists in public life. Furthermore, the testimonies of the activists show that violence is an effective tool capable not only of inflicting pain on women and creating intolerable conflict within their families, but fundamentally of isolating them from their community and derailing their activism, In this context, GBV has become a tool for the advancement of large-scale mining operations.

5.5   CONCLUSIONS

As discussed above, the detrimental impact of large-scale mining on the environment and on rural and indigenous communities in Peru has been the center of attention of a great number of national and international scholars as well as multilateral institutions. In the past two decades, attempts by indigenous movements to influence corporate and state control of natural resources buried within their land have also been examined through many perspectives—including their success at mobilising and securing the support of transnational sectors. In spite of a sustained presence, the contribution of rural and indigenous women to the struggle for rights vis-a-vis the state and mining corporations has been largely neglected by the academic community. A closer examination shows that women's activism did not generate interest among the anti-mining movement either. By focusing on critical moments in the history of women's activism—in particular their unsuccessful efforts to insert themselves within the anti-mining movement—this paper revealed not only the kind of social relations favored by the male-led indigenous movement, but also their leaders' reluctance to place corporate violence against women as an essential part of the discourse against mining operations. Following a historical perspective also helps to highlight that (at least) during the initial stages of their activism, women's lack of agency to present their concerns in terms of national policies meant that they were not able to secure the financial support of transnational organizations. Without human and financial resources, women's community groups were short-lived which in turn reinforced their invisibility and slowed down the success and legitimacy of their concerns.

At the individual level, Margarita and Maxima's reflection on their struggle to be recognized as community leaders provided evidence of gender discrimination and concerted efforts by local groups to erase from the collective conscience of the anti-mining movement that women's activism helped the mining industry to see the need to take on social duties toward women. Furthermore, the activists' testimonies showed that the marginal position of the women's struggle within the anti-mining movement increased their vulnerability to gender-based violence while pushing their precarious social and economic status even further down.

In an attempt to explain the ubiquitous presence of violence in the activists' stories, I reviewed the most important theories examining violence against women. Although a traditional community background—such as the one found in Máxima, Rut, and Margarita's communities—contributes to the creation of an atmosphere that condones GBV, their experiences shed light not only on the tragedy of their personal suffering but more importantly on the complexity of the spider web in which rural and indigenous women activists are trapped. Given this entrenched context of violence, this study argued that it is simplistic to explain the increasing level of violence against women's activists as the result of individual behavior anchored in patriarchal privileges. Instead, it put forward the argument that the violence experienced by activists is fundamentally linked first to the exploitation of natural resources in remote and isolated geographical locations where for centuries indigenous people have lived; and second to the rapid and complex social transformations produced at the community level by the encroachment of resource extraction operations. But that is not all: in spite of ratifying CEDAW and implementing national plans for the protection of women, the way state institutions routinely made allowances for violence against women raised questions regarding the contribution of the Peruvian governments to normalizing violence against women in general, and against activists in particular. Can a state keen to attract global mining investments and secure the establishment of the resource extraction model as the indisputable key to the transformation of Peru into an advanced society remain neutral? The history of the women's testimonies provided evidence that gender norms and gender violence were used by the Peruvian state to restrict and control the advancement of women activists in public life, and that restricting women's activism became a tool that helped to block opposition to large-scale mining operations. The way the police systematically singled them out, imposing on them highly prescriptive norms and procedures that removed the possibility of obtaining justice provided an example of how state institutions erected invisible barriers around women activists. Furthermore, the sustained refusal of the police to be proactive and log the claims made by the activists, coupled with the dismissal of their complaints as inappropriate, becomes an example of how everyday encounters with structural violence are "normalized and rendered invisible through the workings of bureaucratic practice" (Gupta 2012) [46]. For Guzman (2016) [47], failure to log complaints and dismissal of complaints about corporate abuses has become so entrenched that it is responsible for the creation of "spaces of legal marginalization for Cajamarca's activists" (2016 p. 417), through which political participation in Peru is curtailed and (I would add) women's safety compromised. Margarita's sense of dismay at waiting patiently during twenty years for a sentence that never arrived is a concrete example of what Guzman (2016 p. 416) [47] defines as "the systematic archiving of legal claims filed by activists against state and corporate agents," as well as the invisible alliance between the state and mining corporations.

Notwithstanding the struggle of women activists which I witnessed, with the stories of Rut, Margarita, and Maxima I sought to capture aspects of agency and shed light on their quiet heroism and determination to achieve greater control over their lives. It is true that anti-mining activism did not raise their social status within their communities, as happens with male leaders. However, I did offer some marginal evidence that the acquisition of new skills through training and networking brought about personal development and opportunities to re-imagine their lives, grow stronger, develop self-confidence and "critical conscience," a process that among other things encourages "… searching for alternative interpretations of one's situation" (Watkins and Shulman 2010) [91].

5.6   NEWS ARTICLES AND OTHER PUBLICATIONS

[1] Acuña, Máxima (2012). *Testimonio de Máxima Acuña contra la Minera Yanacocha* [Video] YouTube. https://www.youtube.com/watch?v=USk4XOByu48&feature=endscreen 126, 135

[2] Alvarez, S. E. (2000). Translating the Global: Effects of transnational organizing on local feminist discourses and practices in Latin America. *Cadernos de Pesquisa. No 22 Programa de Pos-Graduação em Sociologia e Ciencia Politica*, Universidade Federal de Santa Catarina. 122

[3] Armstrong, R., Baillie, C., Fourie, A., and Rondon, G. (2014). Mining and community engagement in Peru: Communities telling their stories to inform future practice. *IM4DC Action Research*. International Mining for Development Centre of University of Western Australia and Latin America Mining Monitoring Programme. 132, 134

[4] Asaki, B. and Hayes, S. (2011). Leaders, not clients: Grass roots women's groups transforming social protection. *Gender and Development*, 19(2). DOI: 10.1080/13552074.2011.592634.

[5] Bebbington, A., Bury, J., Humphreys Bebbington, D., Lingan, J., Muñoz, J. P., and Scurrah, M. (2008a). Mining and social movements: Struggles over livelihood and rural territorial development in the Andes. *Brooks World Poverty Institute Working Paper no. 33*. https://ssrn.com/abstract=1265582 119

[6] Bebbington, A. and Williams, M. (2008b). Water and mining conflicts in Peru. *Mountain Research and Development*, 28(3):190–195. https://bioone.org/journals/Mountain-Research-and-Development/volume-28/issue-3/mrd.1039/Water-and-Mining-Conflicts-in-Peru/10.1659/mrd.1039.full DOI: 10.1659/mrd.1039. 120

[7] Bebbington, A., Bury, J., Humphreys Bebbington, D., Lingan, J., Muñoz, J. P., and Scurrah, M. (2008a). Mining and social movements: Struggles over livelihood and rural territorial development in the Andes. *Brooks World Poverty Institute Working Paper no. 33*. https://ssrn.com/abstract=1265582

[8] BID Banco Interamericano de Desarrollo (2018). Prevalencia de la violencia contra la mujer entre diferentes grupos étnicos en Perú. A study carried out by J. M. Aguero in collaboration with the Wilson Center. *Nota Tecnica IDB-TN-1455*. 131

[9] Boesten, J. (2010). Inequality, normative violence and liveable life: Judith Butler and peruvian reality. *POLIS Working Papers no 1*. 134

[10] Boesten, J. (2012). The state and violence against women in Peru: Intersecting inequalities and patriarchal rules. *Social Policies*, 19(3). DOI: 10.1093/sp/jxs011. 128, 131

[11] Brock, A. and Dunlap, A. (2018). Normalising corporate counterinsurgency: Engineering consent, managing resistance and greening destruction around the Hambach coal mine and beyond. *Political Geography*, 62:33–47. DOI: 10.1016/j.polgeo.2017.09.018. 134, 135

[12] Burt, J.-M. (2006). Quien habla es terrorista. The political use of fear in Fujimori's Peru. *Latin American Research Review*, 41(3):32–36. 134

[13] Bury, J. (2005). Mining mountains: Neoliberalism, land tenure, livelihoods, and the new peruvian mining industry in Cajamarca. *Environmental and Planning A*, 37:221–239. DOI: 10.1068/a371. 130

[14] Carrington, K., McIntosh, A., and Scott, J. (2010). Globalization, frontier masculinities and violence: Booze, blokes and brawls. *British Journal of Criminology*. DOI: 10.1093/bjc/azq003. 130, 132, 133

[15] Castro, R. and Riquer, F. (2003). Research on violence against women in Latin America: From blind empiricism to theory without data. *Cuadernos de Saúde Pública*. 128

[16] Celendin Libre (2014). Cajamarca: Máxima acuña, heroina ambiental. https://celendinlibre.wordpress.com/2014/10/24/cajamarca-maxima-acuna-heroina-ambiental/ 123

[17] CIEL (2006). Caso12.471-C center for international environmental law. https://www.yumpu.com/es/document/read/24278905/caso-12471-c-the-center-for-international-environmental-law

[18] CIEL (2004). Project update. https://www.ciel.org/project-update/san-mateo-de-huanchor/ 124

[19] CONACAMI (2009). Mujeres victimas de violencia por el estado. https://defensoras.wordpress.com/ 121

[20] Cornwall, A. (2008). Feminist activism for women's rights: What difference does it make? *School of Global Studies*, University of Sussex, UK. https://www.academia.edu/16941306/Feminist_Activism_for_Womens_Rights_What_Difference_does_it_Make 119

[21] De Echave, J., Hoetmer, R., and Palacios Panéz, M. (2009). Minería y territorio en el Perú—conflictos, resistencias y propuestas en tiempos de globalización. http://centroderecursos.cultura.pe/sites/default/files/rb/pdf/mineria%20y%20territorio%20en%20el%20Peru.pdf 119, 120, 121, 122

[22] De Keseredy, W. S. (2019). Intimate violence against rural women: The current state of sociological knowledge. West Virginia University. DOI: 10.18061/1811/87904. 129

[23] Del Aguila, A. (2016). The labour situation of indigenous women in Peru—a study. *International Labour Office*, Geneva. 130

[24] Defensoria del Pueblo (2015). Feminicidio íntimo en el Perú: Análisis de expedientes judiciales (2012–2015). Octavo reporte de la defensoría del pueblo sobre el cumplimiento de la ley de igualdad de oportunidades entre mujeres y hombres. *Serie Informes Defensoriales*, 173–2015-DP. http://peru.unfpa.org/sites/default/files/pub-pdf/Informe-Defensorial-N-173-FEMINICIDIO-INTIMO.pdf 118

[25] Diario NCO (2011). Perú: Declaran estado de sitio en Cajamarca y se estancan las negociaciones con el gobierno. https://diario-nco.com/internacionales/peru-declaran-estado-de-sitio-en-cajamarca-y-se-estancan-las-negociaciones-con-el-gobierno/ 126

[26] Drinot, P. (2011). The meaning of Alan García: Sovereignty and governmentality in neoliberal Peru. *Journal of Latin American Cultural Studies*, 20(2):179–195. 131

[27] Drinot, P. (2014a). Introduction. In Drinot, P. (Ed.), *Peru in Theory*. Palgrave Macmillan. DOI: 10.1057/9781137455260_1. 131, 135

[28] Drinot, P. (2014b). Chapter 8—Foucault in the land of the Incas: Sovereignty and governmentality in neoliberal Peru. In Drinot, P. (Ed.), *Peru in Theory*. Palgrave Macmillan. DOI: 10.1057/9781137455260_1.

[29] Drinot, P. (2011). The meaning of Alan García: Sovereignty and governmentality in neoliberal Peru. *Journal of Latin American Cultural Studies*, 20(2):179–195. DOI: 10.1080/13569325.2011.588514.

[30] EJAtlas Global Atlas of Environmental Justice (2014). https://ejatlas.org/print/san-mateo-de-huanchor-peru

[31] Escobedo, L. (2016). Whiteness in political rhetoric: A discourse analysis of peruvian racial-nationalist "othering." *Studia z Geografii Politycznej i Historycznejtom 5*, pages 257–272. DOI: 10.18778/2300-0562.05.12. 131

[32] Figueroa Romero, D., Rice, R., Estrada, V., and Guimon, S. (2017). Virtual forum violence against indigenous women in the Americas. *Final Summary Report*. Canadian Association for Latin America and Caribbean Studies CLACS. https://poli.ucalgary.ca/manageprofile/sites/poli.ucalgary.ca.manageprofile/files/unitis/publications/1--8355408/CALACS_Virtual_Forum_Violence_against_Indigenous_Women.pdf 117, 128, 129

[33] Facio, A. (2016). Why a gender perspective is needed to analyze the situation of violence against women human rights defenders. *Mesoamerican Initiative of Women Human Rights Defenders*. http://im-defensoras.org/wp-content/uploads/2016/04/286224690-Violence-Against-WHRDs-in-Mesoamerica-2012--2014-Report.pdf 117

[34] Farmer, P. (2004). An anthropology of structural violence. *Current Anthropology*, 45(3). DOI: 10.1086/382250. 123

[35] Figueroa Romero, D., Rice, R., Estrada, V., and Guimon, S. (2017). Virtual forum violence against indigenous women in the Americas. *Final Summary Report*. Canadian Association for Latin America and Caribbean Studies CLACS. https://poli.ucalgary.ca/manageprofile/sites/poli.ucalgary.ca.manageprofile/files/unitis/publications/1--8355408/CALACS_Virtual_Forum_Violence_against_Indigenous_Women.pdf

[36] Flood, M. and Pease, B. (2006). The factors influencing community attitudes in relation to violence against women: A critical review of the literature. *Victorian Health Promotion Foundation (VicHealth)*, Melbourne. 128

[37] Fougere, L. and Bond, S. (2018). Legitimising activism in democracy: A place for antagonism in environmental governance. *Planning Theory*, 17(2):143–169. DOI: 10.1177/1473095216682795. 134

[38] Frente de Defensa Ambiental de Cajamarca (2011). https://fdaccajamarca.blogspot.com/2011/12/solidaridad-con-el-pueblo-de-cajamarca.html 126

[39] Freidenberg, F. and Osornio Guerrero, M. C. (2017). Las consecuencias imprevistas de la participación: La violencia política hacia las mujeres en México instituto de investigaciones jurídicas. Universidad Nacional Autónoma de México. https://www.academia.edu/35846775/Las_consecuencias_imprevistas_de_la_participaci%C3%B3n_la_violencia_pol%C3%ADtica_hacia_las_mujeres_en_M%C3%A9xico 122, 128

[40] Galtung, J. (1969). Violence, peace, and peace research. *Journal of Peace Research*, 6(3):167–191. Sage Publications Ltd. DOI: 10.1177/002234336900600301. 123, 127

[41] Garita Vílchez, A. I. (2013). Nuevas expresiones de criminalidad contra las mujeres en América Latina y el caribe: Un desafío del sistema de justicia en el siglo XXI. Campaña del secretario general de las Naciones Unidas ÚNETE para poner fin a la violencia contra las mujeres. https://studylib.es/doc/5079188/nuevas-expresiones-de-criminalidad 134

[42] Global Witness (2014). The dramatic rise in the killings of environmental and land defenders: 2002–2013. https://www.globalwitness.org/sites/default/files/library/Deadly%20Environment.pdf 118

[43] Grufides (2013). Policía mercenaria al servicio de las empresas mineras. https://issuu.com/grufides/docs/policia_mercenaria_al_servicio_de_1 134

[44] Grufides (2015). http://grufides.org/content/nuevamente-polic-y-trabajadores-de-minera-yanacocha-ingresaron-terreno-de-maxima-acu-y 135

[45] The Guardian (2016). https://www.theguardian.com/sustainable-business/2016/apr/21/peru-farmer-wins-battle-newmont-mining-corporation 125

[46] Gupta, A. (2012). *Red Tape: Bureaucracy, Structural Violence, and Poverty in India*. Duke University Press. 123, 138

[47] Guzman Solano, N. (2016). Struggle from the margins: Juridical processes and entanglements with the Peruvian state in the era of mega-mining. *The Extractive Industries and Society 3*, pages 416–425. DOI: 10.1016/j.exis.2016.02.004. 135, 138

[48] GRUFIDES (2013). Policía mercenaria al servicio de las empresas mineras. *GRUFIDES, APA, CNDDHH, DHSF*. https://issuu.com/grufides/docs/policia_mercenaria_al_servicio_de_1

[49] GRUFIDES (2015). http://grufides.org/content/nuevamente-polic-y-trabajadores-de-minera-yanacocha-ingresaron-terreno-de-maxima-acu-y

[50] Heise, L. L. (1998). Violence against women: An integrated, ecological framework. *Violence Against Women*, 4(3). DOI: 10.1177/1077801298004003002. 128, 129

[51] Horowitz, Lea S. (2017). It shocks me, the place of women: Intersectionality and mining companies' retrogradation of indigenous women in New Caledonia. *Gender Place and Culture*. DOI: 10.1080/0966369x.2017.1387103. 128

[52] INEI Instituto Nacional de Estadísticas e Información (2018). *Censos Nacionales 2017 III Censo de Comunidades Nativas y I Censo de Comunidades Campesinas*, Cuadro no. 3.3 and Cuadro no. 3.30. 129, 130, 131

[53] Jakobsen, H. (2016). How violence constitutes order: Consent, coercion, and censure in Tanzania. *Violence Against Women*. DOI: 10.1177/1077801216678091. 128, 129

[54] Jenkins, K. (2014a). Women, mining and development: An emerging research agenda. *The Extractive Industries and Society*, 2(2). DOI: 10.1016/j.exis.2014.08.004. 117

[55] Jenkins, K. (2014b). *Unearthing Women's Anti-Mining Activism in the Andes: Pachamama and the "Mad Old Women"*. Wiley Online Library. http://onlinelibrary.wiley.com/doi/10.1111/anti.12126/abstract DOI: 10.1111/anti.12126.

[56] Jenkins, K. and Rondon, G. (2015). Eventually the mine will come: Women antimining activists' everyday resilience in opposing resource extraction in the Andes. *Gender and Development*, 23(3):415–431. DOI: 10.1080/13552074.2015.1095560. 122

[57] Kennedy Dalseg, S., Kuokkanen, R., Mills, S., and Simmons, D. (2018). Gendered environmental assessments in the Canadian North: Marginalization of indigenous women and traditional economies. *Northern Review*, 47:135–166. DOI: 10.22584/nr47.2018.007. 120

[58] Lahiri-Dutt, K. (2012). Digging women: Towards a new agenda for feminist critiques of mining. *Gender, Place and Culture*, 19(2):193–212. DOI: 10.1080/0966369x.2011.572433. 122

[59] Li, F. (2009). *Documenting Accountability: Environmental Impact Assessment in a Peruvian Mining Project (PoLAR)*. University of Manitoba. DOI: 10.1111/j.1555-2934.2009.01042.x. 120, 133

[60] Loayza, N. and Rigolini, J. (2016). The local impact of mining on poverty and inequality: Evidence from the commodity boom in Peru. pubdocs.worldbank.org/en/309641458726797039/Peru-Mining-Effects-January-2016.pdf   DOI: 10.1016/j.worlddev.2016.03.005. 119, 120

[61] Lust, J. (2014). Mining in Peru: Indigenous and peasant communities vs. the state and mining capital. *Class, Race and Corporate Power*, 2(3):3. http://digitalcommons.fiu.edu/classracecorporatepower/vol2/iss3/3 DOI: 10.25148/crcp.2.3.16092121. 120

[62] Loayza, N. and Rigolini, J. (2016). The local impact of mining on poverty and inequality: Evidence from the commodity boom in Peru. Last accessed July 25, 2018. DOI: 10.1016/j.worlddev.2016.03.005.

[63] Marti, T. (2012). Indigenous land rights and development in the peruvian Amazon: Communalism versus capitalism. *The Hinckley Journal of Politics*, 13. http://epubs.utah.edu/index.php/HJP/article/view/665/508 131

[64] Menjivar, C. (2008). Violence and women's lives in eastern Guatemala: A conceptual framework. *Latin American Research Review*, 43(3). DOI: 10.1353/lar.0.0054. 129

[65] Mercier, L. and Gier, J. (2007). Reconsidering women and gender in mining. *History Compass*, 5(3). https://doi.org/10.1111/j.1478--0542.2007.00398.x DOI: 10.1111/j.1478-0542.2007.00398.x. 117

[66] Merino Acuña, R. (2015). The politics of extractive governance: Indigenous peoples and socio-environmental conflicts. *The Extractive Industries and Society*, 2(1):85–92. DOI: 10.1016/j.exis.2014.11.007. 120

[67] Mesoamerican Initiative of Women Human Rights Defenders, (2016). Violence against women human rights defenders in Mesoamerica: 2012–2014 Report. http://im-defensoras.org/wp-content/uploads/2016/04/286224690-Violence-Against-WHRDs-in-Mesoamerica-2012-2014-Report.pdf

[68] MIMP, Ministerio de la Mujer y Poblaciones Vulnerables (2014). Boletines y resúmenes estadísticos: Violencia familiar y sexual. http://www.mimp.gob.pe/index.php?option=com_content&view=article&id=1401&Itemid=431

[69] Moser, A. (2004). Happy heterogeneity? Feminism, development, and the grassroots women's movement in Peru. *Feminist Studies*, High Beam Research. www.highbeam.com 120

[70] Neumann, P. (2011). Bureaucracy and legitimacy: A weberian analysis of domestic violence in Peru. *Academia*. https://www.academia.edu/18640941/Bureaucracy_and_Legitimacy_A_Weberian_Analysis_of_Domestic_Violence_in_Peru

[71] Oliver, V., Flicker, S., Danforth, J., Konsmo, E., Wilson, C., Jackson, R., Restoule, J. P., Prentice, T., Larkin, J., and Mitchell, C. (2015). Women are supposed to be the leaders: Intersections of gender, race and colonisation in HIV prevention with indigenous young people. *Culture, Health and Sexuality*, 17(7). DOI: 10.1080/13691058.2015.1009170. 117

[72] Oxfam America (2009). Mining conflicts in Peru: Condition critical. https://www.oxfamamerica.org/static/media/files/mining-conflicts-in-peru-condition-critical.pdf 118

[73] Oxfam America (2011). La Oroya, Peru: Poisoned town. https://www.oxfamamerica.org/explore/stories/la-oroya-peru-poisoned-town/ 121

[74] Pan American Health Organization (2012). Violence against women in Latin America and the Caribbean: A comparative analysis of population-based data from 12 countries. https://www.paho.org/hq/dmdocuments/2014/Violence1.24-WEB-25-febrero-2014.pdf 129

[75] Peru Support Group (2007). Mining and development in Peru: With special reference to the Rio Blanco project. http://www.perusupportgroup.org.uk/files/fckUserFiles/file/FINAL%20-%20Mining%20and%20Development%20inh%20Peru.pdf 118

[76] Red ULAM (2014a). ULAM's submission to the UN commission on the status of women. Peru: Rights of rural and indigenous women endangered by transnational mining. www.redulam.org. Margarita: https://vimeo.com/19033005

[77] Red ULAM (2014b). http://lammp.org/wp-content/uploads/2014/10/Allegation-letter-for-Maxima-Acuna-de-Chaupe.pdf 135

[78] Rénique, G. (2009). Law of the jungle in Peru: Indigenous Amazonian uprising against neoliberalism. *Socialism and Democracy*, 51. DOI: 10.1080/08854300903290835. 133

[79] Rondon, G. (2009). Canadian mining in Latin America: Corporate social responsibility and women's testimonies. *Canadian Woman Studies*, 27(1):89–96. 117

[80] Rondon, G. (2011). Coping strategies used by women living in communities torn by mining activities. *First International Seminar on Social Responsibility in Mining*. https://gecamin.com/srmining/2011/indexd218.html?option=com_content&task=view&id=66 133

[81] Rousseau, S. A. (2015). Paths towards autonomy in indigenous women's movements: Mexico, Peru, Bolivia. *Journal of Latin American Studies*, 48(1):33–60. https://www.cambridge.org/core/journals/journal-of-latin-american-studies/article/paths-towards-autonomy-in-indigenous-womens-movements-mexico-peru-bolivia/AC978259954C07F510585069B2EC89DA DOI: 10.1017/s0022216x15000802. 123

[82] Sanborn, C. and Chonn, V. (2015). Chinese investment in Peru's mining industry: Blessing or curse? *Discussion Paper: Global Economic Governance Initiative*. https://studylib.es/doc/4967552/chinese-investment-in-peru-s-mining-industry--blessing-or... 132

[83] Sanborn, C. and Hurtado, V. (2017). Mining, political settlements and inclusive development in Peru. *ESID Working Paper*, 79. University of Manchester. www.effective-states.org DOI: 10.2139/ssrn.2963665. 133

[84] Svec, J. and Andic, T. (2014). Rethinking empowerment and gender: A case study of domestic violence in Peru. *Population Association of America*. https://paa2014.princeton.edu/papers/141452

[85] Upside-down World (2006). One year since the Bagua massacre—new actors facing a state in crisis in Peru. http://upsidedownworld.org/archives/peru-archives/one-year-since-the-bagua-massacre-new-actors-facing-a-state-in-crisis-in-peru/ 121

[86] UN Expert Group Meeting on Families and Inclusive Societies (2018). Indigenous communities and social inclusion in Latin America. Prepared by Maria Amparo Cruz-Saco and Joanne Toor Cummings. https://www.un.org/development/desa/family/wp-content/uploads/sites/23/2018/05/2--1.pdf 131

[87] UN Special Rapporteur for Indigenous Peoples (2013). Statement on Peru. https://nativenewsonline.net/currents/un-special-rapporteur-issues-statement-peru/ 134

[88] UN WOMEN (n.d). Violence against women: A brief overview of the United Nations and violence against women. *United Nations Entity for Gender Equality and the Empowerment of Women*. https://www.un.org/womenwatch/daw/vaw/v-overview.htm 117

[89] US Geological Survey (2018). Mineral commodity summaries. 118

[90] Vom Hau, M. and Biffi, V. (2014). Mann in the Andes: State infrastructural power and nationalism in Peru. In Drinot, P. (Ed.), *Peru in Theory*. Palgrave Macmillan. DOI: 10.1057/9781137455260_1. 133

[91] Watkins, M. and Shulman, H. (2010). *Towards Psychologies of Liberation*. Palgrave McMilllan. 138

[92] World Health Organization (2005). Multi-country study on women's health and domestic violence against women. Geneva. https://www.who.int/gender/violence/who_multicountry_study/fact_sheets/Peru2.pdf 117, 129

CHAPTER 6

Access to Remedy for Indigenous Communities: A Case Study in Amazonian Peru

Vicki Bilro

6.1   INTRODUCTION

Latin America is one of the most biologically and culturally rich regions in the world, covering a vast expanse of land and home to the world's largest rainforest. Latin America has experienced an economic boom over the last several decades, primarily related to the availability of natural resources and the surge in mining activities. With the already existing challenges of poverty alleviation, increasing education levels as well as job availability and disease prevention in Latin America, there is the added pressure of trying to minimize mining impacts. Mining has environmental, social, and economic effects, some of which reap positive benefits and others which produce negative consequences (Bebbington and Bury, 2013) [3]. The debate continues as to whether the benefits of mining are enough to surpass its impacts and whether there will ever be such a thing as "sustainable mining." However, in light of this consideration, it is especially important to consider indigenous communities in particular as mining frontiers are continuously expanding and moving into remote regions, which are most commonly the homes of indigenous communities (Finer, Jenkins, and Powers 2013) [19].

Indigenous peoples in Peru are strongly impacted by mining activities and have been for many years, especially since the first oil exploration boom in the 1970s (Orta-Martínez and Finer 2010) [41]. There has been a growing number of cases in recent times of indigenous communities organizing protests against the violations of their human rights including their right to land, self-determination, to free, prior and informed consultation, to water, to food, to freedom of expression, and to peaceful assembly and demonstration. The responsibility of these human rights violations lies with the company and State government.

Amid globalization in the 21st century, there has been an increasing emphasis on encouraging companies to recognizing the multiple impacts of business on human rights. Corporate failure to respond to the full negative impact of mining activities on communities can cause financial risks to the company, damage their reputation, produce negative publicity, operational

Table 6.1: Acronyms

Acronym	Description
AIDESEP	Interethnic Association for the Development of the Peruvian Jungle
CNPC	China National Petroleum Corporation
CSR	Corporate Social Responsibility
EMC	Environmental Monitoring Committee
FECONACO	Federation of Native Achuar Communities from the Corrientes River
FECONAT	Federation of Native Kichwa Communities from the Tigre River
FEDIQUEP	Federation of Native Quechua Communities from Pastaza
GM	Grievance Mechanisms
GP	Guiding Principles
HRC	Human Rights Council
ICMM	International Council on Mining and Metals
ILO	International Labor Organization
LAMMP	Latin American Mining Monitoring Programme
MEM	Ministry of Energy and Mines
NGO	Non-Governmental Organization
OEFA	Peru's Agency for Environmental Evaluation and Control
Oxy	Occidental Petroleum Corporation
PRR	Protect, Respect, Remedy
PUINAMUDT	Collaboration of FECONACO, FECONAT, FEDIQUEP, and ACO-DECOSPAT
UN	United Nations
UNDRIP	United Nations Declaration on the Rights of Indigenous People
UNGP	United Nations Guiding Principles

delay as a result of social unrest, divestment campaigns, clean-up costs, legal challenges, as well as loss of license to operate (AmazonWatch, 2011) [2]. Hence, businesses need to operate with a clear understanding of their roles and responsibilities and that of the State in order to identify, prevent and address—at all levels—the adverse human and environmental impacts of their operations.

This project has been commissioned by the Latin American Mining Monitoring Programme (LAMMP 2016) [35], a UK-based charity who work closely with rural and indigenous communities affected by irresponsible and destructive development activities. The focus of this research is on local indigenous communities who are both: victims of human rights and struggling to access remedy and reparations from mining corporations.

6.2   PROBLEM IDENTIFICATION

Under international law, violation of a person's fundamental human rights requires access to effective remedy. This is addressed by Ruggie (2008) [50] as the third pillar of the Protect, Respect, and Remedy Framework (PRR), which states that those whose rights have been violated must be provided access to effective remedy. The United Nations Guiding Principles (UNGP) developed by Ruggie (2011) [51] provides a common platform for States and businesses to operationalize the three pillars of the PRR framework based on an understanding of their relative obligations and responsibilities (Skinner, McCorquodale, and De Schutter 2013) [55].

There are many frameworks and regimes available which address human rights, such as ILO 169 (1989) [29], ISO 26000 (2014) [30], and OECD (2008) [39]; however, Addo (2014) [1] defines the difference between the UNGP to the diverse range of governance and intergovernmental regimes, as its unique mix of governance mechanisms which promote the shared commitment of all stakeholders for effective application. For this reason, the UNGP will become the lens through which this research will be analyzed, with an avid focus on the third of the three pillars.

While the UNGP have revolutionized the idea that corporations have direct human rights responsibilities, as discussed by Cragg, Arnold, and Muchlinski (2012) [11], which challenges previous notions in which corporate social responsibility (CSR) was merely a subject of discussion, critics address limitations in its demarcation between the State's duty to *protect* and the corporation's responsibility to simply *respect* human rights (Murphy and Vives 2013) [38]. Corporations should be expected to do more than respect human rights, but also protect them.

To date, little has been done by corporations regarding the fulfilment of their responsibility to provide access to remedy and as a result victims continue to face human rights violations without any form of recourse or reparation. By focusing on the study of human rights abuses committed over many decades by Pluspetrol in the Peruvian Amazon, this research highlights difficulties encountered by communities due to the lack of operational-level grievance mechanisms. On a more practical level this research also provides recommendations for the implementation of the UNGP.

6.3   RESEARCH OBJECTIVES

Incorporating human rights into company policies and values is no longer solely a moral obligation, it is also a business necessity. Exploring how this obligation is carried out in practice is one of the objective of this research. With this in mind, the study documents difficulties encountered by victims when they attempted to access effective and culturally appropriate remedy as established in the third pillar of the UNGP Framework as well as the contribution of grievance mechanisms (GMs) as part of a formalized process that—although in its infancy—appears to be quite challenging for companies to implement in practice. Consequently, the fundamental aim of this project was to:

- analyze the use and applicability of non-judicial, operational-level grievance mechanisms available for victims who have been subjected to adverse human rights impacts.

It is hoped that this work can be of benefit to communities affected by mining operations in Latin America, and contribute to the search for the most appropriate mechanisms for indigenous communities to obtain access to remedy.

6.4   EXTRACTIVE INDUSTRY IN PERU

The Amazon is the world's largest tropical rainforest. Peru occupies the second largest portion of the Amazon following Brazil. This area of the Peruvian Amazon represents approximately 57% of Peru's total land mass (EY Peru 2014) [17] and it is home to around 60 distinct indigenous groups (Orta-Martínez and Finer 2010) [41]. Peru's abundant supply of natural resources has become an important component of the country's economy, with exports from their extractive sector increasing from 69% in 2000 to 78% in 2011 (Danish Institute for Human Rights 2013) [13] and comprising 4.8% of their GDP in 2013 (EY Peru 2014) [17]. The Ministry of Energy and Mines (2014) [37] declares Peru as the largest silver producer and second largest copper producer in the world. In addition, it is the largest producer of gold, zinc, and steel in Latin America and has significant oil and gas reserves.

Peru was the pioneer oil producing country in Latin America with its first oil well drilled in 1863. Oil exploitation reached the Peruvian Amazon in 1939 by the Ganso Azul Oil Company (Hoy and Taube 1963) [23], with oil production peaking in the 1970s. Exploration at this point was funded by both private companies and the national government. Since the 1990s, the Peruvian State has heavily invested in promoting private investment and free competition by offering attractive mining opportunities for international investors, without competition from nationalized firms. Additionally, price controls were eliminated and there were a number of legal and financial protections for large foreign investors, making Peru an extremely enticing investment opportunity (Bury 2004) [6].

Under Peruvian Law, sub-surface (such as oil) and certain surface (such as forest) resources belong to the State, rather than the private owners of the land on which they thrive (Bebbington and Bury 2013) [3]. This allows the State to bestow mining concessions to companies who are then responsible for reaching an agreement with the communities occupying the land. Historically, most mining companies have been reluctant to provide communities where they have operations with prior consultation and continued communication. In turn, this absence of communication channels has in recent years contributed to social unrest (Finer et al. 2008) [18].

6.5   HISTORY AND POLITICS

Peru, like the majority of countries that make up Latin America, has an extensive and turbulent political history. Peru has alternated between authoritarian and democratic rule upon its independence from the Spanish in 1821. General Velasco—who came to power in 1968 through a military coup—formed the "Revolutionary Government of the Armed Forces" with Nationalist policies (Philip 2013) [42] that alleviated the situation of traditionally excluded sectors. During his time in power, Velasco launched a series of social reforms aimed at enriching the conditions of indigenous peoples. His most notable reforms include the initiation of the national "Day of the Indian" (1969), the legal recognition and organization of Amazonian peoples as "native communities" (1974) and formally recognizing Quechua as a national language with equal status to Spanish (1975), making Peru the first Latin American country to do so (Postero and Zamosc 2004, p. 163) [44]. However, the transition to democracy in 1980, following 12 years of military ruling, saw a period of economic turmoil and rising social tensions (Taft-Morales 2013) [58], alongside the commencement of Peru's civil war which lasted from 1980–2000 (Caston 2013) [8].

Democratic elected governments during the period 1980–1990 were unable to reign in the extremist group *Sendero Luminoso: The Shining Path*, with attacks continuing to escalate. When Alberto Fujimori entered as the new democratic leader of Peru 1990–1995, he commanded severe change by bringing in an aggressive economic reform and stepped up in the use of resilient military tactics to wipe out the *Sendero Luminoso*. His efforts proved successful after the capture of leader Abimael Guzman in 1992, thus effectively "ending" the civil war (Postero and Zamosc 2004, p. 164) [44]. Fujimori's success, however, was built on severe human rights abuses and caused more than 69,000 deaths of which only 35,000 were originally accounted for, as released by the Peruvian Truth and Reconciliation Commission (Knight 2003) [35].

A couple of years after coming into power in 1992 Fujimori's initial political strategies replicated that of previous authoritarian leaders, when he completely dissolved the Peruvian congress and enacted a new constitution in 1993. The new constitution involved significant progress for the rights of indigenous people by declaring the States duty to protect and respect the ethnic and cultural diversity of the nation (Postero and Zamosc 2004, p. 164) [44]. Fujimori was re-elected during the campaign of 1995. However, shortly after being re-elected in 2000, during a visit to Japan Fujimori resigns amidst revelations of electoral fraud and high-level corruption. He was sentenced on April 7, 2009 to 25 years imprisonment for committing "crimes against humanity" on charges of corruption and human rights abuses (Taft-Morales 2013) [58]. Fujimori's trial, arrest, and imprisonment became a landmark legal case and a significant accomplishment for the Peruvian judicial system in terms of endorsing respect for human rights.

Following Fujimori's fall the subsequent presidents formed a period of relative political stability, economic growth, and poverty reduction. Ollanta Humala came into power in 2011 as a result of putting at the center of his presidential campaign the formation of a new relationship with the mining industry, incorporating demanding higher royalties from mining companies and reducing inequality and poverty (Kaczmarski 2012) [33]. Humala won significant support from impoverished indigenous voters who have been negatively affected by the mining boom. One of his first acts as head of State was to sign into effect the Law of the Right to Prior Consultation for Indigenous and Native Peoples, which led Peru to be the first Latin American country to integrate International Labor Organization Convention 169 (ILO 169) into national legislation. While this was a significant step in the right direction for Peru, his agenda met challenges; one major stumbling block was defining who is considered Indigenous (Sanborn and Paredes 2014) [53] and the other was his refusal to adopt a radical agenda colluding instead with traditional economic and political interests (Sanborn and Paredes 2015) [52].

6.6   SOCIAL CONFLICT

Increase in foreign investment since the 1990s provided a boost for Peru's economy, however its effects on indigenous communities are still felt today: many communities have been displaced, productive agricultural lands have been reduced in size while many plots are considered infertile and water sources have been seized for mining interests or are now contaminated (Postero and Zamosc 2004) [44]. These violations of basic human rights are part of a more extensive list, which are not being respected and upheld by companies as well as the Peruvian government.

Situations of social unrest have become all too common in Peru, increasing as the number of mining projects escalates (Hodge 2014) [22]. There has been a significant increase in the number of conflicts from 2006 to the end of 2009, the year in which the historic case in Bagua occurred which caused 34 deaths. Of the 249 conflicts recorded up to April 2012, 149 were associated with mining activities (OPM and ICMM 2013) [40]. Conflicts relating directly to mining activities have many causes; from dissatisfaction with the minimal benefits that local communities receive from mining revenue, the lack of capacity, and political will of the Peruvian government to regulate the industry, manage local conflicts and redress grievances to serious mistakes made by mining companies refusing to address grievances and concerns of local communities (Slack 2009) [56].

With the increasing number of social conflicts in Peru, often governments came to power with the explicit agenda of ensuring communities benefit from mining and improving social inclusion (OPM and ICMM 2013) [40], however this is a difficult task to implement and maintain. For example, an article published by Jamasmie (2015) [31], a specialist in CSR and the Latin American Market, exposed President Kuczynaki's bias in favor of the mining industry. In Jamasmie's view, Kuczynaki's decision to declare martial law in parts of Peru's southern highlands torn by anti-mining protests ignored real communities concerns about serious health environmental caused by "Las Bambas" copper mine. The suspension of political rights during a 30 day state of emergency period gave police permission to enter houses without search warrants. Although the protest against "Las Bambas" mine was portrayed as merely one example of social conflict that has augmented the delay of $21.5 billion worth of mining projects in Peru in recent years (Jamasmie 2015) [31] the decision to declare martial law went against the fundamental aims of the government's initial development plan.

6.7   SIGNIFICANCE OF THE RESEARCH

Mining is a long-term business activity and as such subject to complex (often unforeseen) costs. Eager to avoid unnecessary costs companies are always keen to identify cheaper alternative that allow them to mitigate social and environmental risks rather than engage in expensive remediating practices. For example, Janine Ferretti, head of Inter-American Development Bank's environmental division, estimates that generally around one percent of a company's total costs is required to mitigate social and environmental risks (The Economist 2016) [59].

This research explores two case studies through the lens of the UNGP with a view to establish how useful and practical GMs can be to communities seeking remedy. Communities' understanding of the importance of GMs is crucial because when mining operations negatively impact local communities, the UNGP expects companies to have mechanisms in place that offer communities redress and remedy.

6.8   METHODOLOGY

This research project encompassed a thorough archival data collection of secondary literature of both formal and informal sources. A diverse range of literature from varying perspectives enabled an unbiased approach and provided the basis for understanding GMs and the critical role their absence played in the chosen case study, which was then analyzed through the framework.

6.8.1   DATA COLLECTION

Secondary data analyzed through this research was collected from company reports, peer-reviewed articles and community reports. Table 6.2 lists the main sources and types of data that were employed for the relevant topics, however please note this is not a complete list and refer to the References for the collection in its entirety.

6.9   THEORETICAL FRAMEWORK

6.9.1   DEVELOPMENT OF THE UNITED NATIONS GUIDING PRINCIPLES

Promoting respect for human rights is a core purpose of the United Nations (UN) and, therefore, the institution has effectively developed a number of important resources. For example following the continual injustices perpetrated against indigenous peoples, the UN Declaration on the Rights of Indigenous Peoples was developed (UN General Assembly 2007) [60]. The following year, the Human Rights Council implemented the UN "Protect, Respect, and Remedy" (PRR) framework focused directly on the impact of business on human rights. This was

Table 6.2: **Data types and sources used for analysis**

TOPIC	TYPE	SOURCE
Grievance mechanisms	Peer-reviewed articles	Corporate Social Responsibility Initiative at the John F. Kennedy School of Government
	Company reports	Centre for Social Responsibility in Mining (CSRM) at the University of Queensland
	Company report	Global Oil and Gas Industry Association for Environmental and Social Issues (IPIECA)
	Company report	Office of the Compliance Advisor/Ombudsman (CAO)
	Company report	International Council on Mining and Metals (ICMM)
	Peer-reviewed articles	Centre for Research on Multinational Corporations
	Community and company reports	Business and Human Rights Resource Centre
Case study	Peer-reviewed articles	Subterranean Struggles: New Dynamics of Mining, Oil, and Gas in Latin America
	Peer-reviewed articles	Business and Human Rights: Indigenous Peoples' Experiences with Access to Remedy
	Peer-reviewed articles	Oil frontiers and indigenous resistance in the Peruvian Amazon
	Community reports	Amazon Watch
	Community reports	The Economist
	Community reports	Cultural Survival
	Community reports	London Mining Network
	Community reports	Alianza Arkana
	Community reports	Chaikuni Institute
	Company reports	Pluspetrol
	Company reports	Rio Tinto

later endorsed by a set of Guiding Principles on Business and Human Rights in 2011 to operationalize the three pillars of the framework. Ruggie (2008) [50] defined the three pillars that govern the framework as:

Pillar 1: The State's duty to protect human rights abuse by third parties through appropriate policies, regulation, and adjudication;

Pillar 2: The corporate responsibility to respect human rights by encouraging business enterprises to act with due diligence; and

Pillar 3: Access to effective and culturally appropriate remedy from both States and corporations, of judicial and non-judicial form.

The UN Framework and its Guidance Pillars (GPs) have come under a great deal of scrutiny since their initiation. Although they received praise for combining the commitment of all key stakeholders to affirm the importance of non-state actors as discussed by Addo (2014) [1], they have been criticized for making the distinction between the State's duty to *protect* and the corporation's responsibility to *respect* human rights (ICMM 2009b [25]; Murphy and Vives 2013 [38]). Cragg (2012) [10] suggested that these boundaries be removed to expand the obligations of businesses beyond the responsibility to simply respect human rights, but to also protect them.

Huijstee, Ricco, and Ceresna-Chaturvedi (2012) [24] acknowledge that the UNGP are the most authoritative and internationally recognized framework incorporating business and human rights, yet they do not create international legal obligations that can be enforced for companies. It is important to note that prior to the implementation of this framework and GP, corporations had no direct accountability of their human rights responsibilities. Ruggie (2008) [50] openly writes in the PRR framework that there is no single silver bullet for the issues in the business and human rights domain, instead all those involved must learn to do things differently. Steps are being taken to continually move forward in what has been and will undoubtedly continue to be a lengthy process to achieving protection and respect of the human rights of all people.

The Third Pillar and Understanding Grievance Mechanisms

It remains today that there are still corporate-related human rights abuses without the provision of remedy and reparation. The third pillar of the framework manages the provision of adequate access to remedy for victims who have been subjected to human rights abuse as a result of business activities. The term grievance is employed in the UNGP to describe the injustice of an individual's or group's sense of entitlement, which may be based on law, contract, promises, customary practice, or general notions of fairness (Ruggie 2011 [51]; SOMO 2014 [57]). The GP follow on to utilize the term GM to indicate any methodical, State-based or non-State-based, judicial or non-judicial process to address and remedy grievances concerning business-related human rights abuse.

Principle 25 of the UNGP—the foundational principle—states that State-based judicial or non-judicial GM should form the basis of a wider system of remedy, which is supported by operational-level GM and supplemented by collaborative initiatives. Table 6.3 demonstrates the breakdown of mechanisms made available by the UNGP and each component of the figure is substantiated by operational principles 26–30. As highlighted by the red box in Table 6.3, the focus of this research is on operational-level GM which is explained by Principle 29. This principle states:

> *To make it possible for grievances to be addressed early and remediated directly, business enterprises should establish or participate in effective operational-level grievance mechanisms for individuals and communities who may be adversely impacted.*

Wildau et al. (2008) [61] define operational-level GM as a practical and controlled approach for companies to accept, investigate, respond to, and remedy grievances experienced by local communities, without the correspondence of the courts and in a timely manner. Legal remedies come under the domain of the State and are generally the most appropriate avenue in developed countries whose legal systems are fully functioning with fast and effective legal procedures. However, considering some of the worst human rights abuses occur in countries where the rule of law is ineffective or completely absent, or where legal procedures are slow and resource intensive, it is important to have other avenues for those affected to pursue (Business and Human Rights Initiative 2010) [7]. Judicial and non-judicial mechanisms should collaborate to minimize the burden on either mechanism and achieve the optimal outcome for those affected.

The main focus of companies should be to identify and address grievances early before they escalate. Paying attention to human rights from a business perspective should be more than merely complying with the law. Most company value statements incorporate principles such as "integrity," "honesty," and "respect," which are similar to the core values of the UN Framework and form the basis of most human rights frameworks. In addition to moral boundaries, there are financial and legal considerations which can benefit from human rights due diligence. Company value can increase and business opportunities can arise as a result of socially aware trading (Business and Human Rights Initiative 2010) [7]. Addressing the gap between societal and business goals s the first step toward responsible mining operations. Accordingly, the focus of this research is on mining companies and the role they should play in providing access to remedy for local communities.

6.10   DATA ANALYSIS

The choice of UNGP as the analytical tool was for its relationship between business and human rights, allowing the focus to be on how companies can address and provide access to remedy and reparation for local indigenous peoples affected by their mining activities. By grasping a thorough understanding of operational-level GM and studying Pluspetrol's approach, it was clearly evident that the company had failed to implement GM as a method for providing remedy at the community level. Further research into how other companies tackle GM in practice led us to choose Rio Tinto and its La Granja mine as the secondary case study. The analysis of GM utilized by Rio at La Granja served as basis to recommendations which could facilitate communities impacted by Pluspetrol access to remedy.

6.10.1   CASE STUDY: PLUSPETROL

History of the Mining Concession

The Oil Block 1AB mining concession—focus of this research—has over the years been owned by a succession of companies. It is located in the Department of Loreto, which is situated in the north-eastern corner of Peru and covers approximately one third of the country's territory. In spite of its vast extension, it is the most sparsely populated region due to its remote location in the Amazon rainforest, yet it is abundant in natural resources. This concession resides on three important rivers, all of which are tributaries of the Amazon River and home to a number of indigenous communities.

Mining activities in this region affect the indigenous Quechua people of the river Pastaza, the Achuar, Quechua and Urarinas of the Corrientes River and the Kichwas of the Tigre River (Doyle 2015) [14].

There is a long history of mining exploitation behind the Oil Block 1AB. The name as such was given in 1971 by the founding company, US-based Occidental Petroleum Corporation (Oxy). (Clancy and Kerremans 2015) [9]. Pluspetrol briefly shared Oil Block 192 with Oxy until they secured a 15-year concession in May 2000 from the Peruvian government. The concession was renamed Oil Block 192 in 2012 after the block size was increased.

As explained by Miller (2015) [36], Oil Block 192 has extreme importance for the Peruvian State due to its significant oil production and large oil reserves that still remain, as well as its future potential to integrate with other blocks located over the oil corridor. Oil Block 192 is the largest oil field in the Peruvian Amazon and produces nearly 20% of the country's total oil output (Sanborn and Paredes 2015) [52]. In spite of significant health and environmental problems, Sanders (2015) [54] considers that the oil output of the concession is a major factor behind the decision of the Peruvian governments to push to renew or re-sell the concession upon expiration in August 2015. Affected communities used this date as an opportunity to address their long-standing demands regarding the use of their land and territories (Doyle 2015) [14]. They demanded an environmental assessment, reparation and due compensation for use of their lands, legal land titling and a genuine consultation and participation process in accordance with the ILO 169 (see Appendix B). The communities' greatest concern was that Pluspetrol would walk away freely when the concession ended without addressing their demands; something they had experienced once before with Oxy in 2000.

Understanding all events that have led up to the expiration of Oil Block 192 is important and are documented in tabular form in Appendix A.

Who is Pluspetrol?

Pluspetrol—Peru's biggest oil group—is a subsidiary of Argentine oil company Pluspetrol S.A and China National Petroleum Corporation. In 2003 Pluspetrol created a "strategic alliance" with China National Petroleum Corporation (CNPC), one of the world's largest oil companies. Although CNPC purchased 45% of the concession in 2003, they earn themselves the label as the "silent partner" due to their effectively invisible participation in the operations. Pluspetrol is controlled by the Dutch holding company Pluspetrol Resources Corporation N.V (Bnamericas).

Pluspetrol operates in Latin America—across Peru, Colombia, Argentina, Venezuela, and Bolivia—along with exploratory activities in Africa (Clancy and Kerremans, 2015) [9]. Pluspetrol usually engages smaller companies to carry out many of their practical operations on the ground which makes it difficult to understand who is ultimately responsible.

Pluspetrol Corporate Legacy of Environmental Destruction

Shortly after Occidental Petroleum Corporation (Oxy) started its large-scale production of nearly 40% of Peru's oil, concerns about the environmental impact of Oil Block 192 began to surface. Although the area had been described as the "most polluted region in the country" by the government's National Office for Natural Resource Assessment in 1984 (Bebbington and Bury 2013) [3] the Peruvian State's response was subdued. This was partially due to the lack of regulatory norms as the country's first framework for environmental protection was adopted in 1990 and the Hydrocarbon Act was introduced in 1993 (Orta-Martínez and Finer 2010) [41].

Oxy's production operations ended in 1999 and transferred to Pluspetrol. Although Pluspetrol assumed all environmental and social responsibilities of the predecessor when they agreed to take over the concession, the company remained involved in protracted legal discussions and persistently denied any responsibility over any environmental damage produced by previous occupants of the oil block.

Given this situation in 2007, the Achuar people of the Corrientes River filed a lawsuit against Oxy demanding compensation and remediation of the damage left from the daily dumping of toxic produced waters during several decades. Three years later, the case won an appeal to proceed to the U.S. Federal Court and a mutual settlement of the claims in the litigation was reached in 2015. The success of this litigation highlights the importance of external bodies such as the indigenous federations and international human rights institutions, who worked closely with local communities to obtain remedy.

It is important to stress that since Pluspetrols occupation of Oil Block 192 in 2000, the company was besieged by local indigenous communities for its continual oil contamination. Analysis of Pluspetrol's mining activities in Oil Block 192 showed serious environmental degradation, which are estimated date back to the time when operations started and led to human rights violations. For example over the course of four years from 2008–2012, more than 100 crude oil spills were identified in Oil Block 192.

Although local communities protested for the absence of an environmental impact assessment it wasn't until 2012 that the Peruvian government launched a multi-sectoral commission to investigate the exact environmental state of the region in which Pluspetrol operated.

The first results were released in January 2013 and validated evidence presented by local communities : water quality tests revealed that the Pastaza River had 352 times the allowed levels of heavy metals and hydrocarbons. The following results in August 2013 showed the Corrientes River to be in an alarmingly similar state to that of the Pastaza, and was soon announced to be in a state of environmental emergency. Finally, in November 2013 the Tigre river results were released and the same situation followed (Kistler and Alianza Arkana 2013) [34].

That same government study exposed that all but two people (out of a group of 199) from the affected communities had blood-cadmium levels critically above average (Sanders 2015) [54]. In addition, people from these communities presented tumors, skin ailments, chronic illnesses, miscarriages, increased blood pressure, muscle debilitation, nervous system damage, to mention but a few (Doyle 2015) [14].

Despite efforts, the great majority of health and environmental concerns of the affected communities are still yet to be addressed. Indigenous people continue to use contaminated sites. Their lives as well as local wildlife continues to be threatened by irresponsible oil exploitation. Given the importance of Oil Block 192 the government agreed to formal dialogues between the Ministry of Energy and Mines and the Indigenous Federations. Cultural Survival (2015) [12] reported that no agreement was reached during the process of consultation.

For this research understanding the companies and the complexity of their operations was a vital process in analyzing their impact on Oil Block 192. Pluspetrol has other mining concessions in Latin America which have followed a similar path to that of Oil Block 192, hence the importance of learning lessons from its operations in Peru.

The Role of Indigenous Peoples of the Peruvian Amazon

In spite of a serious lack of State presence and corporate responsibility the work of the different indigenous organizations representing the affected people was characterized by significant co-ordination (Doyle 2015) [14]. Different groups joined efforts and worked together to achieve important milestones in their situation. Repeatedly these groups stressed that monetary gain was not their primary interest as their greatest concern was remediation of their land before new exploitation started. The relevant Indigenous Federations include:

- FECONACO: Federation of Native Achuar Communities from the Corrientes River

- FECONAT: Federation of Native Kichwa Communities from the Tigre River

- FEDIQUEP: Federation of Native Quechua Communities from Pastaza

6.10.2   PLUSPETROL'S FAILURE TO IMPLEMENT GRIEVANCE MECHANISM (GM)

UNGP literature (Appendix C. Principle 31) recommends that companies implementing GM should follow specific standards in order to ensure that members of the community trust mechanisms and are able to use them. ICMM (2015) [28] adds that companies should aim to establish the mechanism as early on in the process as possible, make it culturally appropriate, accessible, predictable, equitable, transparent, rights-compatible, as well as a source of continuous learning and based on engagement and dialogue. Consequently, when defining, designing, implementing and monitoring GM, corporations must follow the criteria defined in Principle 31 of the UNGP to ensure their effectiveness in practice.

Since the establishment of its mining operations, Pluspetrol has had a severe impact on the Quechua, Achuar, Urarinas, and Kichwa indigenos peoples. In violation of ILO 169 these communities have not once been involved in prior consultation. They are unaware of their rights and their concerns have barely been addressed since 1971. Achievements they have made thus far have been heavily dependent on NGOs which have presented the grievances of all four groups as one. From the GM point of view while the four indigenous communities may have common requirements for reparation of damage, they also have cultural differences and varied preferences for grievances that had not been addressed. Fulton et al. (2015) [20] warn about debilitating external factors such as the dependence of indigenous communities on NGO support for accessibility to mechanisms. In the case of Pluspetrol, the ambiguity surrounding the term *remedy* and varying definitions of people's expectations of remedy with those of the company exacerbate the frustration of indigenous people with the process of seeking redress which in turn led them to accept what was offered to them, thus settling for the perception that "something is better than nothing." Fulton recommends that when addressing grievances it is essential to recognize the varied ethnic and socio-economic backgrounds of the aggrieved parties. Applying one solution will not meet the needs or requirements of all involved, thus causing inequalities in remedy implementation.

Moving Forward for the Local Communities of Oil Block 192

One of the biggest issues facing the indigenous communities impacted by Pluspetrol is the lack of consultation since the commencement of operations. In order to positively move forward, Pluspetrol (or the next mining company that takes over the concession) need to engage with the communities so that a repeat of the last two concession incidences does not occur. This research has provided evidence that communities have been suffering the impacts of Pluspetrol activities, some of which are irreversible, without seeing any form of benefit. There needs to be improved health facilities for those incapacitated by health problems from mining contamination, increased employment opportunities for people to support their families and remediation of contaminated land and water so everyone has access to clean drinking water, to mention but a few.

It shouldn't take a third concession renewal for the local communities to learn what their rights are. They are slowly making progress to lessen the impact of what has occurred throughout their region and to obtain compensation from Pluspetrol and Oxy, but it has been at the cost of 45 years of degradation and destruction of their human rights. If from the beginning communities had been made aware of what their legal rights are and how to prevent being abused by mining companies, they may have been able to save themselves years of heartache and irreparable damage.

The local communities have managed to obtain some form of remediation throughout the course of the mining activities, mainly through the formation of indigenous organizations. The organizations have mobilized together, coordinated outreach, lodged complaints to the State and sourced international mechanisms. These organized movements have achieved significant milestones in this case for the affected peoples of the three rivers. The difficulties encountered by the communities affected by Pluspetrol illustrates the importance of joint efforts and together lobby for the implementation of operational-level GM from Pluspetrol or the next mining company to assume the concession.

It is unfortunate that the Peruvian State has remained relatively silent over the course of the activities of Pluspetrol. As the first mining concession in the Peruvian Amazon, Oil Block 192 has been a learning curve for the indigenous communities, State, companies and external bodies, such as NGOs. In order to move forward, lessons must be learned and drawn on from all parties involved to ensure it will not happen in the future. Since this concession is such an enormous resource, producing nearly 20% of Peru's total hydrocarbon output, it will surely be renewed. Whatever decision is made in the near future needs to address the local community requirements, all of which can potentially be remediated through the use of non-judicial GM.

6.10.3 CASE STUDY: RIO TINTO AND ITS LA GRANJA MINE

Pluspetrols lack of utilization of the UNGP and implementation of GM steered this research to investigate mining companies that have implemented the GP to avoid human rights abuse and/or address adverse impacts of their mining activities. While it is well-documented that Rio Tinto has been responsible for human rights abuses during some of their business operations (see Richard (2010) [47] for examples); in its "La Granja project" the company made a real attempt to implement many non-judicial GM making it an important learning component for the future of Oil Block 192.

Although still in its exploration stage, "La Granja"—considered among one of the most important mines in the world—has a long history and over the years has had three corporate owners. The period between 1993 and 1999 when La Granja was the property of Canadian company Cambior is associated with collusion between the Peruvian state and the corporation to implement an aggressive programme of land acquisition and relocation of the communities (Flynn and Verhara 2015) [21]. Infrastructure developments and resettlement activities carried out during this period by Cambior played a key role not only in transforming the villages but also in fueling resentment among local farmers (Castillo and Brereton 2018). It is not surprising that when Rio Tinto became the owner in 2005 the company faced strong opposition from local communities (Environmental Justice ATLAS) located within the area of influence of the mine. Tensions escalated until in 2008 claiming that they had not been consulted directly (JBS Market Research 2017) [32] communities forced Rio Tinto to shut down temporarily its operations (REUTERS 2008) [46]. By 2009, Rio Tinto had a thorough understanding of the challenges it faced and what the population feared (Flynn and Verhara 2015) [21]. It is interesting to note that local communities affected by Pluspetrol and local communities impacted by Rio Tinto "La Granja" project shared similar list of demands they felt should be addressed. The four main elements of these requirements are:

1. compensation for damage;

2. conduct of prior consultation in accordance with ILO 169;

3. participation in environmental monitoring programs; and

4. environmental assessment and remediation of damages caused.

The difference between the two business approaches here discussed is that each of the issues listed above are issues that Rio Tinto confronted and addressed in their work. At La Granja, Rio Tinto established a sustainable development program that encompassed three focus areas for the local communities: Education, Health, and Productive Development. According to Rio Tinto these were guided by the principles of dialogue, transparency, responsibility, and commitment; reflective of the UNGP (Rio Tinto 2014) [49].

The educational component of the development programme contributed to the training of 176 pre-primary, primary, and secondary school teachers within the district. Regarding health development the company provided more than 6,000 services in comprehensive health, ophthalmological and dental campaigns, as well as assisted training for health care professionals and donated equipment for more than 20 health establishment in the district. In productive development, the company aimed to make subsistence farming activities a beneficial opportunity for families and established projects such as a guinea pig breeding project that benefited 430 families. Through these initiatives, the communities are receiving long-term benefits rather than monetary compensation.

Rio Tinto state that they also conducted prior consultation and during community consultations engaged with locals thus setting the basis for a trusted relationship. Rio Tinto worked with local communities to establish an Environmental Monitoring Committee (EMC) and as part of this initiative the company contracted Futuro Sostenible, a Peruvian NGO, to train volunteer community members in monitoring and environmental management putting emphasis in measuring water quality, which during consultations had been established as a major community concern.

During these training members also learn how La Granja's exploration operations function and perform monthly visits with full access to all areas, with a view to develop a formal report which identifies concerns and is sent to HSE. The EMC program is expected to provide community members with training which in turn they can use to monitor the impact of mining activities.

The mechanisms employed by Rio Tinto do not suggest that the La Granja project has been simple and straightforward since its initiation. It is instead an indication of how the implementation of GM can provide remedy to communities who have expressed issues regarding compensation for damage, prior consultation, and environmental monitoring, while benefiting both the communities and company in the short- and long-term. These GM align with the effectiveness criteria by being transparent through provision of regular communication about project progress, thus contributing to the process of building trust among stakeholders in accordance with the UNGP and ILO 169.

6.11   CONCLUSIONS

Mining activities have an exhaustive number of impacts on local communities and those impacts can and should be expected to cause grievances. Those grievances need to be identified and remediated effectively and efficiently to not only administer the impacts of those affected, but to benefit the company's business operations and reputation. GM, as explicated in the UNGP, are an important tool for responsible companies to prevent escalation and facilitate the resolution of grievances experienced by local communities and other stakeholders. Although the UNGP do not create international legal obligations that can be enforced for companies, they are the most authoritative and internationally recognized framework incorporating business and human rights and hence, they were chosen as the analytical framework for this research.

Common themes that emerged from local communities regarding the usability of grievance mechanisms were the right to prior consultation, the lack of representation of all those involved, the deficiency of understanding their legal rights and the dependence on non-governmental organizations. By mitigating these issues from the commencement of mining activities or adopting GM to remedy issues that arise, companies can avert complications for themselves and the local communities. Rio Tinto has done so by engaging in consultation with community members of the project at La Granja mine. In addition, they established initiatives such as sustainable development programmes addressing health, education, and production and engaged environmental monitoring committees made up of locals who are trained in monitoring and managing the environment. These GM encompass the effectiveness criteria defined by the UNGP and have minimized social conflicts and enhanced production capabilities.

6.12   FURTHER RESEARCH

The adverse impacts of mining operations on human rights have been evident since its commencement, however it is only in recent years that this issue has become widely publicized and important for corporations and governments. As mining activities are moving into more remote regions to access extensive reserves, indigenous communities and their resources are being severely impacted. This is becoming increasingly common in the Amazon basin, where local communities are subjected to the negative effects of mining without being provided forms of remedy or reparation.

Research showed that there is abundant information on designing, implementing and monitoring grievance mechanisms, which is decidedly simpler in theory to accomplish than in actual practice. This research project was a stepping stone in exploring the useability of grievance mechanisms in real-world situations in Peru. There is further work to be done in investigating case studies of similar situations and analyzing the success or failure of certain grievance mechanisms in providing remedy. Analyses need to be collated so lessons can be learned from every experience in order to improve the usability and applicability of grievance mechanisms.

6.13   APPENDIX A: TIMELINE OF PLUSPETROL CASE STUDY

The timeline of events since the commencement of Oil Block 1AB in 1971 (Clancy and Kerremans 2015 [9]; Kistler and Alianza Arkana 2013 [34]) in Table 6.3.

6.14   APPENDIX B: DEMANDS OF THE AFFECTED COMMUNITIES

Requirements from the affected communities prior to consultation of the concession renewal (from Doyle 2015) [14] are:

1. the conduct of an environmental assessment and the remediation of environmental damage;

2. clean-up and land titling;

3. compensation for the use of their lands and damages caused by oil activity over the course of the last 40 years;

4. the conduct of prior consultation and participations processes in accordance with the form and conditions established under ILO Convention 169;

5. recognition of community environmental monitoring systems;

6. the participation of indigenous organizations in the development, evaluation, and monitoring of environmental management instruments; and

Table 6.3: Timeline of events

Date	Event
1971	Oil operations begin in Oil Block 1AB by Occidental Petroleum Corporation.
1996	Pluspetrol commences operations in Peru.
2000	Pluspetrol take over Oil Block 1AB.
2003	Pluspetrol creates a "strategic alliance" with CNPC, who purchase 45% of the concession.
2007	FEDIQUEP initiate independent indigenous territorial monitoring program in the Pastaza river basin.
2011	Four indigenous federations affected by Pluspetrol's activities unite together to create the organization PUINAMUDT, to defend their traditional territories and demand justice for oil pollution on their land.
May, 2012	Peaceful mobilization of indigenous people in the Pastaza to demand a stop to the impunity of oil contamination and action from the Government.
June, 2012	In response to the pressure, multi-sectoral commission is formed by Peruvian government to investigate contamination from oil activities in affected river basins.
Oct., 2012	First entry of different technical government bodies to take soil and water samples in Oil Block 1AB.
Jan., 2013	Results from Pastaza are published, exposing life-threatening levels of heavy metals and hydrocarbon in soil and water sources.
Mar., 2013	Environmental emergency declared for the Pastaza River Basin.
Aug, 2013	Environmental emergency is declared for the Corrientes River.
Nov., 2013	OEFA fines Pluspetrol with USD 7 million for irreparable environmental damage and "disappearing" Lake Shanshococha in Oil Block 1AB in the Pastaza.
Nov., 2013	Environmental emergency is declared for the Tigre River.
June, 2014	Mobilization by Quechua people of the Pastaza, since they were told that solutions would be found by the multi-sectoral Commission.
Jan.–Feb., 2015	Kichwa people block the river Tigre demanding prior consultation, compensation, and reparation.
Feb., 2015	Blocking of oil wells by the Achuar communities demanding reparation for damages caused.
Aug., 2015	Expiration of concession for Oil Block 192.

7. transparency, damage assessment, and sanctioning of those responsible for pollution.

6.15 APPENDIX C: PRINCIPLE 31 EFFECTIVENESS CRITERIA

The Effectiveness Criteria for Non-Judicial Grievance Mechanisms as defined by Ruggie (2011) [51]:

1. *Legitimate*: enabling trust from the stakeholder groups for whose use they are intended, and being accountable for the fair conduct of grievance processes;

2. *Accessible*: being known to all stakeholder groups for whose use they are intended, and providing adequate assistance for those who may face particular barriers to access;

3. *Predictable*: providing a clear and known procedure with an indicative time frame for each stage, and clarity on the types of process and outcome available and means of monitoring implementation;

4. *Equitable*: seeking to ensure that aggrieved parties have reasonable access to sources of information, advice, and expertise necessary to engage in a grievance process on fair, informed, and respectful terms;

5. *Transparent*: keeping parties to a grievance informed about its progress, and providing sufficient information about the mechanism's performance to build confidence in its effectiveness and meet any public interest at stake;

6. *Rights-compatible*: ensuring that outcomes and remedies accord with internationally recognized human rights; and

7. *A source of continuous learning*: drawing on relevant measures to identify lessons for improving the mechanism and preventing future grievances and harms;

   Operational-level mechanisms should also be:

8. *Based on engagement and dialogue*: consulting the stakeholder groups for whose use they are intended on their design and performance, and focusing on dialogue as the means to address and resolve grievances.

6.16 NEWS ARTICLES AND OTHER PUBLICATIONS

[1] Addo, M. K. (2014). The reality of the United Nations guiding principles on business and human rights. *Human Rights Law Review*, 14(1):133–147. DOI: 10.1093/hrlr/ngt041. 149, 155

[2] AmazonWatch (2011). The right to decide. http://amazonwatch.org/assets/files/fpic-the-right-to-decide.pdfRetrievedfromhttp://amazonwatch.org/assets/files/fpic-the-right-to-decide.pdf 148, 174

[3] Bebbington, A. and Bury, J. (2013). *Subterranean Struggles: New Dynamics of Mining, Oil, and Gas in Latin America*. Austin, TX, University of Texas Press. 147, 150, 158

[4] Bnamericas. Pluspetrol Norte company profile. https://www.bnamericas.com/company-profile/en/pluspetrol-norte-sa-ppn.

[5] Pluspetrol   S.A.   https://www.bnamericas.com/company-profile/en/pluspetrol-sa-pluspetrol-argentina

[6] Bury, J. (2004). Livelihoods in transition: Transnational gold mining operations and local change in Cajamarca, Peru. *The Geographical Journal*, 170(1):78–91. DOI: 10.1111/j.0016-7398.2004.05042.x. 150

[7] Business and Human Rights Initiative (2010). How to do business with respect for human rights: A guidance tool for companies. *The Hague: Global Compact Network Netherlands*. 156

[8] Caston, E. (2013). Civil war in Peru (1980–2000). https://modernlatinamericanart.wordpress.com/2013/05/08/civil-war-in-peru-1980-2000/ 151

[9] Clancy, C. and Kerremans, S. (2015). *Oil Dependency and the Peruvian Amazon*. Peru, Chaikuni Institute. 157, 158, 164

[10] Cragg, W. (2012). Ethics, enlightened self-interest, and the corporate responsibility to respect human rights: A critical look at the justificatory foundations of the UN framework. *Business Ethics Quarterly*, 22(1):9–36. DOI: 10.5840/beq20122213. 155

[11] Cragg, W., Arnold, D. G., and Muchlinski, P. (2012). Guest editors' introduction human rights and business. *Business Ethics Quarterly*, 22(1):1–7. DOI: 10.5840/beq20122212. 149

[12] Cultural Survival (2015). Peru concludes consultation on lot 192 before agreement reached with indigenous federations. https://www.culturalsurvival.org/news/peru-concludes-consultation-lot-192-agreement-reached-indigenous-federations 159

[13] Danish Institute for Human Rights (2013). Peru: Human Rights and Business Country Guide. 150

[14] Doyle, C. (Ed.) (2015). Business and human rights: Indigenous peoples' experiences with access to remedy. Case studies from Africa, Asia, and Latin America. Chang Mai, Madrid, Copenhagen, AIPP, Almaciga, IWGIA. 157, 159, 164

[15] EarthRights International (2015). Maynas v. Occidental. https://www.earthrights.org/legal/maynas-v-occidental

[16] Environmental Justice ATLAS. https://ejatlas.org/conflict/la-granja-rio-tinto

[17] EY Peru (2014). Peru's mining and metals investment guide: 2014–2015. Lima, EY. 150

[18] Finer, M., Jenkins, C. N., Pimm, S. L., Keane, B., and Ross, C. (2008). Oil and gas projects in the Western Amazon: Threats to wilderness, biodiversity, and indigenous peoples. *PLoS One*, 3(8):1–9. DOI: 10.1371/journal.pone.0002932. 150

[19] Finer, M., Jenkins, C. N., and Powers, B. (2013). Potential of best practice to reduce impacts from oil and gas projects in the Amazon. *PLoS One*, 8(5):1–14. DOI: 10.1371/journal.pone.0063022. 147

[20] Fulton, T., Ha, J., Karimian, M., Lerner, E., Meier, A. C., and Plessis, I. (2015). What is remedy for corporate human rights abuses? Listening to community voices, a field report. New York, SIPA Print. 160

[21] Flynn, S. and Vergara, L. (2015). Land access and resettlement planning at La Granja, *CSRM Occasional Papers: Mining-Induced Displacement and Resettlement Series*. Eds., Deanna Kemp and John Owen, University of Queensland. 161, 162

[22] Hodge, R. A. (2014). Mining company performance and community conflict: Moving beyond a seeming paradox. *Journal of Cleaner Production*, 84:27–33. DOI: 10.1016/j.jclepro.2014.09.007. 152

[23] Hoy, D. R. and Taube, S. A. (1963). Power resources of Peru. *Geographical Review*, 53(4):580–594. DOI: 10.2307/212387. 150

[24] Huijstee, M. V., Ricco, V., and Ceresna-Chaturvedi, L. (2012). How to use the UN guiding principles on business and human rights in company research and advocacy: A guide for civil society organisations. Netherlands, SOMO, CEDHA and Cividep India. 155

[25] ICMM (2009a). Human rights in the mining and metals industry: Handling and resolving local level concerns and grievances. London, International Council on Mining and Metals. 155

[26] ICMM (2009b). Human rights in the mining and metals industry: Overview, management approach and issues. London, International Council on Mining and Metals.

[27] ICMM. Rio Tinto Minera Peru: La Granja project. https://www.icmm.com/en-gb/case-studies/rio-tinto-minera-peru-la-granja

[28] ICMM (2015). *Good Practice Guide: Indigenous Peoples and Mining*, 2nd ed., London, International Council on Mining and Metals. 160

[29] Convention concerning Indigeous and Tribal Peoples in Independent Countries (ILO No. 169), (1989). https://www.ilo.org/dyn/normlex/en/f?p=NORMLEXPUB:12200:0:: NO::P12100_ILO_CODE:C169 149

[30] ISO (2014). ISO 26000. *Guidance on Social Responsibility.* 149

[31] Jamasmie, C. (2015). Over $21bn worth of mining projects delayed in Peru due to social conflict. http://www.mining.com/over-21bn-worth-of-mining-projects-delayed-in-peru-due-to-social-conflict/ 152, 153

[32] JSB Market Research (2017). https://www.jsbmarketresearch.com/construction/rio-la-granja-copper-mine-development-cajamarca-project-profile 162

[33] Kaczmarski, M. (2012). Comment: Viewpoint—Peru's corporate and social inclusion pursuit—Luis Miguel Castilla. *The Banker.* http://search.proquest.com/docview/1033312424/fulltext/CBFA47EC4837416BPQ/10?accountid=14681 152

[34] Kistler, S. and Arkana, A. (2013). The misleading art of commercial deception: How to transform a major polluter into an environmentally sustainable and socially responsible company. http://alianzaarkana.org/blog/2013/12/09/the-misleading-art-of-commercial-deception-or-how-to-transform-a-major-polluter-into-an-environmentally-sustainable-and-socially-responsible-company/ 159, 164

[35] Knight, J. (2003). Statistical model leaves Peru counting the cost of civil war. *Nature*, 425(6). LAMMP (2016). Mission statement. http://lammp.org/about-2/mission-statement/ DOI: 10.1038/425006b. 148, 151

[36] Miller, A. E. (2015). Asserting indigenous rights in Peru's plagued and prolific block 192. http://amazonwatch.org/news/2015/0730-asserting-indigenous-rights-in-perus-plagued-prolific-block-192 157

[37] Ministry of Energy and Mines (2014). Peru: Mining country. http://www.minem.gob.pe/_detalle.php?idSector=1&idTitular=159&idMenu=sub149&idCat 150

[38] Murphy, M. and Vives, J. (2013). Perceptions of justice and the human rights protect, respect, and remedy framework. *Journal of Business Ethics*, 116(4):781–797. DOI: 10.1007/s10551-013-1821-0. 149, 155

[39] OECD (2008). OECD Guidelines for Multinational Enterprises. 149

[40] OPM and ICMM (2013). Responsible mining in Peru: International council on mining and metals. 152

[41] Orta-Martínez, M. and Finer, M. (2010). Oil frontiers and indigenous resistance in the Peruvian Amazon. *Ecological Economics*, 70(2):207–218. DOI: 10.1016/j.ecolecon.2010.04.022. 147, 150, 158

[42] Philip, G. (2013). Nationalism and the rise of Peru's General Velasco. *Bulletin of Latin American Research*, 32(3):279–293. DOI: 10.1111/blar.12031. 151

[43] Pluspetrol (2012). Purpose, vision and values. http://www.pluspetrol.net/e-proposito.html

[44] Postero, N. and Zamosc, L. (Eds.) (2004). *The Struggle for Indigenous Rights in Latin America*. Sussex, Academic Press. 151, 152

[45] Rees, C. (2008). Grievance mechanisms for business and human rights: Strengths, weaknesses and gaps. *Working Paper No. 40*. Cambridge, MA, John F. Kennedy School of Government, Harvard University.

[46] REUTERS (2008). Rio Tinto suspends Peru construction after protest. https://www.reuters.com/article/us-metals-peru-riotinto/rio-tinto-suspends-peru-construction-after-protest-idUSTRE49J7LE20081020 162

[47] Richard (2010). Rio Tinto: A shameful history of human and labour rights abuses and environmental degradation around the globe. http://londonminingnetwork.org/2010/04/rio-tinto-a-shameful-history-of-human-and-labour-rights-abuses-and-environmental-degradation-around-the-globe/ 161

[48] Rio Tinto (2013). Why human rights matter: A resource guide for integrating human rights into communities and social performance work at Rio Tinto. Australia and UK, Rio Tinto plc and Rio Tinto Limited.

[49] Rio Tinto (2014). La granja project—CEO water mandate. https://ceowatermandate.org/files/lima/O'Keeffe_(Lima2014).pdf 162

[50] Ruggie, J. G. (2008). Protect, respect and remedy: A Framework for business and human rights. *UN Doc A/HRC/8/5*. DOI: 10.1162/itgg.2008.3.2.189. 149, 155

[51] Ruggie, J. G. (2011). Guiding principles on business and human rights: Implementing the United Nations Protect, Respect, and Remedy, framework. *UN Doc A/HRC/17/31*. 149, 155, 166

[52] Sanborn, C. and Paredes, A. (2015). Getting it right? Challenges to prior consultation in Peru. *Centre for Social Responsibility in Mining, Occasional Paper Series*, pages 1–26. 152, 157

[53] Sanborn, C. and Paredes, Á. (2014). Case study: Peru. *Americas Quarterly*, pages 54–60. 152

[54] Sanders, E. (2015). Peru's eviction of oil company could set precedent. http://www. culturalsurvival.org/news/perus-eviction-oil-company-could-set-precedent 157, 159

[55] Skinner, G., McCorquodale, R., and De Schutter, O. (2013). The third pillar: Access to judicial remedies for human rights violations by transnational business: ICAR, CORE, ECCJ. 149

[56] Slack, K. (2009). *Mining Conflicts in Peru: Condition Critical*. Oxfam America. 152

[57] SOMO (2014). The patchwork of non-judicial grievance mechanisms: Addressing the limitations of the current landscape. Netherlands, Centre for Research on Multinational Corporations. 155

[58] Taft-Morales, M. (2013). Peru in brief: Political and economic conditions and relations with the United States. *Congressional Research Service*. 151

[59] The Economist (2016). From conflict to co-operation. http://www.economist.com/ news/americas/21690100-big-miners-have-better-record-their-critics-claim-it-up-governments-balance 153

[60] United Nations Declaration on the Rights of Indigenous Peoples: Resolution/Adopted by the General Assembly (2007). 153

[61] Wildau, S., Atkins, D., Moore, C., and O'Neill, E. (2008). *A Guide to Designing and Implementing Grievance Mechanisms for Development Projects*. Washington DC, The Office of the Compliance Advisor/Ombudsman. 156

CHAPTER 7

Translating Values into Action: What Can Be Done?

Rita Armstrong, Caroline Baillie, Andy Fourie, and Glevys Rondon

7.1 INTRODUCTION

If one accepts that, for the time being anyway, mining must continue (despite its unsustainable outlook) this book demonstrates the enormous changes required in the mining sector, and the educational projects and courses needed to develop a new generation of engineering students, who are willing to make these changes to create an industry which is more equitable and just. As one small example of potential positive changes, this chapter builds on the work of Chapter 2, and attempts to draw together the views of different stakeholders in that context, with the aim of creating some guidelines for change. Chapter 2 presented the perspectives of a selection of community members at two mine sites about their experience of social conflict with the Peruvian state and with mining companies. These detailed narratives revealed deep hostility toward the state and the mining companies based upon treatment by police and by company officials. The experiences of these men and women are varied but have given rise to a common belief that neither the government nor the companies hold any respect for the values, beliefs and needs of community members.

There may be community members at those mine sites who do not share those views, and it is also evident that each company has a different social history with the communities that live within or alongside areas affected by mining. However, there is enough evidence from research in the same region and from other sites of social conflict in Peru to acknowledge that negative perceptions of mining companies and government bodies are common, are based on similar issues, and that the depth of these feelings are evident in the strength of opposition to mining in many parts of Peru. The results of the first project were presented as a set of guidelines which laid out the key areas for improvement needed to create equitable negotiations in the future. This second project aimed to deepen the context of these guidelines by exploring how companies and Governments who have positive relations with communities, appear to have applied them, or from the perspective of the communities, could do so in the future.

This research project entailed the following.

1. Interviewing senior employees of mining companies based in the Andes; academics who have worked as consultants or researchers on the mining industry; government representatives from the Ministry of Mining and National Office of Dialogue and Sustainability; representatives from community-based groups in areas impacted by mining; farmers and NGOs. Details of these interviews, and an explanation of their codes is set out in Appendix A.

2. Formulating suggestions to enact the guiding principles in order to facilitate the development of a mutual understanding of company-community interests.

3. Creating two short films to be used as a pedagogical tool in teaching and learning about the impact of mining on communities.

It is not the aim of the chapter[1] to provide a management tool for either government or companies but to provide some meaningful direction about what needs to change in the values and behavior of companies, governments, communities and NGOs to develop more equitable approaches to community engagement in the future. The use of narrative in both Chapter 2 and in this chapter aims to "develop translation and mediation tools to help make visible the difference of interests, access, (and) power" (Fischer 2003, p. 3) [2] of those people at the center of mining economies.

7.2   COMPARATIVE VIEWS: THE STATE, COMPANIES, COMMUNITIES, AND NGOS

7.2.1   HOW THE STATE IS PERCEIVED

It is very clear, from Chapter 2, that community members—including farmers, urban residents, and local members of community-based organizations—believed the state was working to protect the interests of mining companies even when those companies had caused environmental damage or had not engaged with communities in an open and democratic way. One of the other outcomes of that research was that it was not clear what local people meant when they talked about "the state": negative feelings about the state were most often specifically associated with police actions at social protests and attitudes of prosecutors who failed to take their claims of unfair treatment seriously, and generally with feeling that the state had not protected them.

In this section we compare the different perceptions of the "state" from each of the group of interviewees listed above, while attempting to disentangle which state institution is being discussed.

---

[1] This chapter is largely based on a report published for and supported by the IM4DC (International Mining for development center UWA, Australia.

7.2.2   ABSENCE OF INSTITUTIONAL REPRESENTATION, OF REGULATION

A persistent theme in interviews with companies, community-based groups, and government itself is the absence of government in the remote regions of Peru. However "absence" means different things to each group.

For companies, absence reflects a lack of capacity to govern at the regional level, and a lack of capacity to deliver development programs to local people even when monies are available to do so, particularly from the canon.

Senior Personnel in Mining Companies

"C2: Governments in developing countries is invariably absent from the more remote areas and indeed even in the areas where you do find government present it is so often staffed with people who are unprepared, perhaps even unwilling, to take the steps that are necessary to take advantage of the development that the investment will bring. But without that involvement of government, it's a project that will not be sustainable. It will not be fully accepted either by the agency that has the ultimate responsibility for operating it, or the next mayor of the town who is anxious to discredit all of the work of his predecessor."

"C3: And the other thing that you see is there is very little presence. The government has no presence basically where we operate. So most of the areas where we operate are very remote, have very low indicators in terms of poverty, nutrition, education and so forth … We need more presence of the government and not just for the sake of the mining industry, or for any other industry, but for the sake of being more inclusive."

Anthropologist

A1: The capacity of the local level to manage public funds is very, very limited. This is not good for the people. The communities want the government to protect them. They want now for the companies to step away, and say the governmentneeds to respond. Now the government is trying to feel how to come to that. It is complicated. There are too many bodies and each one thinks differently and the co-ordination between them is very weak.

A few company people blamed corruption at the regional level for the rise of social conflict.

C3: no wonder they are frustrated. Regional governments in Cajamarca, Ancash, Tumbeh, Piura are managed by corrupt regional governments.

Company 4: When you ask the people who are in the regional government, the ones who have complained about mining, when you ask "what have you done," "what have you done for the people." Nothing. They haven't done anything.

Beyond protesting. Fine, you can do your protest, you can be against the mining operations in Peru but what have you done for them with all the money, all the resources that came into

Cajamarca? Absolutely nothing. So you have the worst poverty indicators in the country, the worst extreme poverty indicators of Peru in Cajamarca. With millions and millions of dollars and you have leaders such as Santos, and you ask the question—is this really protecting the environment the answer for the people of Cajamarca, the rural communities who have no nutrition, no education. Those things you ask yourself is the really way forward.

Anthropologists on the hand perceived local government as having a lack of capacity:

A2: The government transferred money to the regions without any preparation for the management of these funds. The lack of capacity of the local government to manage these funds increase the risk connected to social conflict because people didn't see the benefit of the money. Even the rules about how to spend the money are strange. Local people who are expected to invest the money in infrastructure don't have the capacity to hire people with technical knowledge. That situation is strange. Even in this case about how to invest in a proper project, the state just put some regulations and don't give proper courses for capacity building.

A representative of one of the large companies believes that the government wants the benefit of mining revenue but is unwilling to fulfill its regulatory obligations in return. It is worth noting that, in Peru, large-scale mining is monitored by the national government but smaller projects are managed by regional authorities.

C4: the company gets there, the company starts engaging with the community, they start engaging with the local authorities and it's not until two years or one and a half years that I start my permitting process to explore, that the government starts auditing me every three months and they don't come. When do they come? At the next stage when I need an EIA approval for my exploration activities when I need to start workshops with the communities, they have to be present. And what they do sometimes, they delegate this to the regional authorities.

C4: Who audits the large-scale mining in Peru? It's not the regional government, it's the national government, the central government. And the problem is that they are not by my side, and they are losing credibility and they are losing the leadership that they should have around the mining sector … Auditing. You know, taking care of the environment, making sure that mining companies are mitigating properly any kind of impact that they are producing. If things are right and they are performing well, well you need to tell that and communicate that properly. If mining companies are doing things wrong, ok the same way. Explain what is going on, what's happening. Have these meetings with the communities every three months and update the communities around the mining, with what is going on in the mining sector, what's going on with mining in the district… So don't wait until a mining company gets there and announces that there is a discovery or announces some activity for the government to go "oh, let's go!"

7.2.3   ABSENCE OF PROTECTION

Community-based organizations and farmers, on the other hand, feel the absence of protection in the face of excessive police brutality during protests.

Female members of community-based women's organization against mining

W2: considering everything, carried on with the states of emergencies, and the people have realized now, because of so much injustice… now for us there are no laws, there is no one to protect us, everything is an abuse

W3: we feel like we are orphans, because there is no one that we can complain to, they don't listen to us, for them we don't exist

W3: because the one who should be condemned, is not condemned but on the contrary, our laws are upside down, from the central government, we feel weak in that aspect, there is not support, there is no justice for the poor, that is what we say

W4: they are abusing us, they don't think that we are human beings and that we have been left marginalized, scared of the state which put in a state of Emergency.

Farmer from Celendin district

F5: there is no authority here in Cajamarca, in Celendin that really exercises justice like it should be, they use the law like they want to, according to how they act, according to the decisions that they make, and they don't use the law, the rules like they should do, for justice for those who need it the most, for the poor who are demanding their rights…

NGOs, which seem to fulfill the role of the state in regional areas (in terms of legal advice, information about mining), feel that it should be possible for the state to be both regulator of mining and protector of the people:

N2: the state should have a intermediary perspective, a perspective, let's say, a perspective that looks to protect the community, that looks to lend services, and when, and the companies should have a determined role, one that is to adjust itself to the conditions given to them, by the state and by the community, but the reality in Peru, what is currently happening, is that because there is a current urgent need to have strong investment, from these companies, the state has opened up, are being as flexible as they can be with the guidelines and norms, and the companies are taking advantage of that.

Most academics and two government employees also agreed that the state had failed in its role to protect rural and Indigenous populations. One of them said "*The communities, frankly, don't trust the government. The government didn't do good work before*" but qualified this by stating that government was now attempting to rectify that situation by having established the National Office for Dialogue and Sustainability, and by having the Ombudsman[2] record the level of social conflict at mine sites around Peru.

7.2.4 THE STATE DOES NOT VALUE THE ATTACHMENT TO LAND

Linked in with the feeling of not being "seen" by the state, is the feeling that the state does not recognize, or give equal value to, the culture of local people. Again, we hear from the women's group against mining:

> W3: *if we look at the countries here, Ecuador, Colombia, Venezuela, and Chile, they have the same problems as we do, the state doesn't understand the importance that our territories have to us the people, the state doesn't understand the importance that those mountains have to us, because for us they are important, right? and that is something the state also, doesn't value and neither does the company, they come here and say that the mountain is not worth anything, but for me the mountain is important, and with my mountain I do my land, I do my rituals for my cultivations, and the state doesn't understand that.*

> W3: *we are talking about mining and the government always talk about a responsible mining, and that we can live from gold, but that is false, I think that the best development is to have water, the best source of development, otherwise we wouldn't be able to be drinking this tea at this moment, for me mining is not compatible with agriculture, nor with water, for me the best development of a town, is the water and I think that, and I think that we will be able to resist, right?*

7.2.5 THE STATE COLLUDES WITH COMPANIES

There is general agreement that, due to the history of mining in Peru, local people have little basis on which to trust either the state or the company. But while companies now see the state as not fulfilling its duties as regulator or developer in regional areas, some communities see active collusion with companies who have breached environmental regulations and human rights.

7.2.6 WOMENS' GROUP AGAINST MINING

> W6: *the state doesn't see the people, as Peruvian people, but they govern for the transnational's, in favor of them and not in favor of our country, it is a problem that when we protest, the state itself sends the police and sends them to throw gas bombs, and so many abuses toward people, I don't know, maybe we have hope that some day we will have a government that governs us for us, it is for us, that sees the people, for our Peru, for our country.*

---

[2]The *Defensoria del Pueblo* was constitutionally formed in 1992; the first *Defensor* was appointed in 1996. The Ombudsman's, as it is known in English, is an autonomous national body.

M7: maybe we would have to first activate a system to first guarantee that there is independence on behalf of the state, and for the state to become a guarantor of the people and so that would allow, for example, that the state takes on the role of overseer of the mining companies, and also the role to guarantee respect of the human rights of its people, that is currently not happening because the state is almost forced to be the co-partner of these companies, it is a partner of the companies and the state sees itself, well I think also that this is because of the economic dynamics that we currently have and the system that we have, in this neo-liberal system, unfortunately it has given way for these issues, because of themes of corporate governance.

A representative from one of the smaller mining companies, a Peruvian sociologist, commented on the power dynamics of the state-company relationship:

C5b: conflicts show that companies have more power than the villagers because the companies are supported by the government because they are the source of taxes and source of investment (and corruption), so the government try to make the projects work. So there is a kind of inequality in the power structure because the villagers don't have enough support from the government to be in equal condition to translate all the issues to the mining companies. Once we have this inequality on the power relations, mining companies doesn't feel that—they don't have to be, to have to make any negotiations because they will demand to the government to employ the force to restate the situation and avoid the issues because they are the source of the taxes and investments and they require from the government that they attack the communities to keep the things ongoing.

7.2.7   ENVIRONMENTAL IMPACT ASSESSMENTS SHOULD BE ACCESSIBLE AND EASY TO UNDERSTAND

Many community members claimed, in the previous report that it was difficult to understand EIAs, and there was not enough time for consultation. The Ministry of Energy and Mines (MEM) has made what it believes are progressive changes to gaining community endorsement for an Environmental Impact Assessments. Public hearings for EIAs were only heard in Lima until 2003; after this modification they could be heard in regional towns. Furthermore, EIAs are now available on the internet. Despite these changes, power imbalances remain. Fabiana Li (2009) [4] has recorded instances of only partial community representation in town hall meetings in Cajamarca, for example; furthermore not everyone has access to the internet and even when town people have access to the internet, the language and length of the EIAs have remained the same.

W5: we don't have access and the information is so complex, it's so big, that no one is reading the one thousand four hundred pages which on top of that has technical terms, and with regards to the social licence of which Lourdes talked about, right? the community according to the law, they have a right to be consulted, but that is if the country fulfilled the laws.

The EIA to which this woman is referring comprised a 530-page "Technical Component" and a 130-page "Social Component, as well as numerous appendices with additional maps, figures, survey results, interview guides, and other data" (Li, Fabiana, 2009, p. 222) [4].

7.2.8   THE OFICINA NACIONAL DE DIALOGO Y SOSTENIBILIDAD (ONDS) IS A SIGN OF IMPROVEMENT

New institutional channels have been created to resolve social conflict. These include the *mesas de diálogo* (dialogue roundtables) and *mesas de desarrollo* (development roundtables) established by regional authorities and by the *Oficina Nacional de Diálogo y Sostenibilidad* (National Office of Dialogue and Sustainability or ONDS). None of the company representatives mentioned the ONDS or said that they had made use of the Development Roundtables.

Nonetheless they, and some academics, acknowledged that the government had made improvements in addressing social conflict.

> *C4: there is goodwill and things have changed and improved a little bit, I have to be honest about that.*

> *A1: To be honest, they have been changing the framework, now they are trying to build a system where they have let's say a theoretical framework to explain why the need to be there in order to create a process of dialogue. But in my opinion that has been paying off, and working—that is I am not talking about how good or bad each process is. But they have been able to reduce the violence in some cases and to keep the possibility of a project open. The strategy from the government, which is something which the people demand, is that the government needs to be responding.*

The obstacles facing the ONDS are, however, significant. The Development Roundtables requires representatives from all relevant ministries and this is difficult to achieve (physically) let alone reach a consensus. The Commissioners from the ONDS have to liaise primarily with the Ministry of the Interior, but a Development Roundtable requires representatives from the government departments that deal with water, agriculture, health, development, etc. In the two years since the office was created, it has dealt with over a 100 cases. As one Commissioner from the ONDS stated:

> *G4: each Government Minister has their centralized office—and their speciality is really technical. For example we have the national water authority. Or the office for the evaluation of the environment (OEFA). And these institutions have done really good work evaluating and monitoring the activities of mining or oil projects.*

> *Generally when we get information, then we need to meet with these institutions. We need to understand all the elements of the controversy and then we program a meeting with the company or the community. In my case, in some cases that I have, the first step*

is to get information, the second step is get a meeting with the national institutions of the government or maybe regional gov or local gov depending on the characteristics of the conflict. Then we have to program a meeting with the community. One thing is the technical information but another aspect of the conflict is the perception of the community and it is really important for us to get that perception, and to understand what they think about the project. This is the process.

The two Commissioners whom we interviewed also acknowledged that the communities do not trust the government but felt that improvements were being made:

G4: This table is working to build trust with the Indigenous people. It's like a window for the community. The communities, frankly, don't trust the government. The government didn't do good work before and this government (Ollanta Humala's government) has responded to the questions of these communities.

An NGO representative also commented on the difficulty of getting representatives of all 18 ministries to a Dialogue Roundtable:

N2: (speaking about the Amazon) All the indigenous gathered, they are still gathering, and they have been requesting a Dialogue Table for about two years now, only this year it was established, yes it was this year, then the indigenous leaders said that they wanted to hold the dialogue in their communities, and well their communities are far, really far, to get there you have to go in a boat on the river for ten hours, so then the PCM intelligently said that that was ok, that they will have dialogues in their communities, so then the leaders organized various dialogue spaces, erm, but well, they proposed that they meet in an area of Iquitos, a city in the jungle, and so all of the communities went with their groups but the state didn't show up.

The community-based organizations and farmers, on the other hand, did not mention the ONDS at all, either in the first or second round of interviews. This could be because they are not interested in a "dialogue table" but want the mining in their area to cease altogether. Furthermore a lot of rural people may be unaware of the ONDS because at the time of our research it had only recently formed and did not have a strong regional presence.

One anthropologist also felt that some movement had been made to strengthen the presence of the government in the regions but cited similar obstacles to improvements in other government sectors, i.e., working against centrist tendencies and also getting ministries to cooperate with each other.

A1: The capacity of the local level to manage public funds is very, very limited. This is not good for the people. The communities want the government to protect them. They want now for the companies to step away, and say the government needs to respond. Now the government is trying to feel how to come to that. It is complicated. There are too many bodies and each one thinks differently and the co-ordination between them is very weak.

An employee of the Ministry of Mines did not feel that Roundtables were the ultimate or final solution but were a way of avoiding violence and mediating conflict.

7.3   HOW MINING COMPANIES ARE PERCEIVED

7.3.1   A BAD HISTORY IN PERU

There is general consensus that mining in Peru has a negative history; this is not unusual in a global perspective where most mining companies, until the late 1990s—which coincided with the rise in environmentalism, in Indigenous activism, and a famous court case against BHPBilliton by the Yonggom people of Papua New Guinea—paid little attention to the environmental and social impact of mining.

One anthropologist who takes an interest in the historical development of mining in Peru charts a course from the days of "old mining":

> *A3: In the 50s it was national kind of companies. But also you have a middle sized companies, Peruvian important ones like Buenaventura, Hoschilds. In these times there was a kind of a culture about how to relate with, how to deal with the social issues. And that culture was part of our society, completely biased. So the idea was that the people from mining were oligarchs and the rest were Indians, Indigenous people who have no value at all. These companies came and tried to negotiate with the local communities in very asymmetrical ways with the idea that "well you want something, I give you some money or some job for one or two and yes please" (snaps fingers) "go out from the land." "If you don't want that, I'm gonna kill you, that's all." It's my right. Like a conqueror you know.*

The period of nationalized mining was no better, in his opinion. He believes that when the government nationalized large sections of the mining industry, the explicit social contract was: "I am going to, you know, contaminate everything but I am going to give you all jobs and services." This, he says, was partly successful because mine workers became politicized and the problem then was not about contamination but about fair remuneration for work, and a better system of government. The move to liberalization, and the growth of foreign companies who then espoused "corporate social responsibility" seemed to promise a better deal for communities.

> *Anthropologist A3: Because they reproduce this old culture, bad culture, everybody—people, community, media—they started to see a contradiction because you have all of these companies saying "we're following international standards, following whatever, but on the land we behave as usual." So there started to be conflicts, huge conflicts, it was very bad because at the beginning when these companies first came, they started to say to the communities "oh you know we have different standards, we have changed, this is the new mining" but at the end of the day it was the same as usual. I remember the communities had a lot of hope "oh the new mining it's coming and we're going to have electricity and we're going to have services and we're going to be all happy and safe because …." They no longer say that*

in any place in Peru but this was the mood. After this first encounter, now everybody say "I don't want the mine" because the promise of "new mining" was betrayed.

These views are shared by company people. Here is a lengthy quote from the CEO of an exploration company in the southern Andes which encapsulates the same views;

C6a: Peru was the richest gold country in South America accessible for Europe and the Spanish. So a lot of slavery started in Peru and the Peruvian population was forced to work up in the mines. Definitely with time, the country started civilizing itself somehow but nevertheless the Inca families, or our Indians, managed to keep on with their traditional activities which was mining, agriculture, ok. By the time when modernization came in, the Indian families kept on with their activities and strong families, strong Peruvian family with capital started involving themselves in mining business. In due course, taking advantage of the position of the Peruvian people, they abuse the situation regarding the non-respect to salaries, the non-respect to humans and the impossibility to give them better opportunities, forced them to work for the eight to ten oligarchy families of Peru. So somehow slavery and colonialism had the same meaning at the end of the day.

With foreign capital, in the beginning the story—and I am talking 20, 25 years ago—in the beginning, the story was nice. There were not much community conflicts. The conflicts really started when the development of mining started booming. People with not enough cultural education between (about) human relations were taking care of projects. Somehow a lot of promises were given to people, to Peruvian people, who owned the superficial territories that had communities, and these promises were never complied. A lot of—some of these mining companies, especially talking about Peruvian mining companies started threatening very much the surface areas, started contaminating, started misusing the waters, and basically extending their capital interests without leaving any benefit to the communities.

It is also significant that all the academics and company interviewees agreed that one of the downfalls of the so-called "new mining" in the 1990s was to delegate the management of community relations to Peruvian partner companies whose staff were from Lima, were generally wealthy and who had little regard for rural people in the Andes.

Anthropologist A3: And that means that some of these medium sized miners from Peru they make alliance with these huge mining companies, in a minority role. Like you have a share, say 10%, %50 whatever with the international company. But at the beginning the people from Peru, the miners from Peru say "I know how to deal with my own people so I am going to be in charge of the social issues." So for the first time in the 90s it was the minority partner from Peru who took responsibility for social issues and you wonder what happened?! What did they reproduce? The old strategies.

7.3.2   ARROGANCE: A LEGACY OF THE PAST

The previous report gave many examples of why community people perceived the behavior of company officials as arrogant. Here are similar views put forward by company people, as well as NGOs.

> C2: *As I said earlier on, sometimes the Peruvians are the bigger problem. And Peru is an interesting example in that over such a short time it has expanded rapidly not just economically but in terms of education. But what hasn't necessarily changed have been the unspoken class system that has existed here in Peru since the Spanish I am sure. And you get engineers and geologists who may have come up out of the community but who then believe they are one step, if not more, above the farmers. And they act it. And that just makes me mad. They are so arrogant. They are so arrogant.*

> C3: *There was a lot of arrogance and lack of knowledge from our industry.*

> C6b: *Lima, traditionally, hasn't cared or respected, or really understood or taken the time to. Lima has its wealth, has its wealthy people, its hierarchies and they're arrogant. And the terrorism situation that happened, the suffering that took place out in the Andes in the jungle, it was despicable. But it wasn't until the Sendero Luminoso started to put bombs in Lima, in Miraflores, that anything happened, that anyone paid any attention and that's a little bit the problem.*

The last quote about attitudes toward rural people in the Andes is particularly significant. As a consequence of the conflict between Shining Path (*Sendero Luminoso)* and the armed forces in the 1980s and 1990s, 69,280 Peruvians died and disappeared. The Peruvian Truth and Reconciliation Commission (CVR) concluded that 46% of these can be attributed to Shining Path, 24% to another terrorist group, MRTA, and 30% to the army and police forces (Boesten 2012) [1]. The CVR also concluded that 79% of the people affected by the conflict lived in rural areas and that 75% spoke Quechua or another indigenous language: "The tragedy suffered by rural communities, from the Andes and from the jungle, Quechua and Ashaninka, farmers, poor, and uneducated was not felt by the rest of the country," the report says; it also laid bare the "root causes of Peru's conflict: inequality and racism. Many Peruvians still might not want to admit that, but no one denies it either. The divide between rich, cosmopolitan Lima and the poor, rural regions discussed in the report still seems impossible to erase" (ICTJ n.d.) [3].

The prevailing arrogance in the attitude of foreign and local exploration or mining staff is exemplified here:

> C6b: *previously exploration companies would just come in and walk around do what they want. A lot companies, a lot of junior companies just exploring. They traditionally have not always given enough to the communities or have surface rights agreements they just wander around, they wander in there before they make any agreements. Our policy is that*

you don't set foot on, anywhere, until you know who are the traditional owners. That's from the initiation of this company you don't set foot onto someone else's ... it's like I don't walk into your house and open up your fridge uninvited ... (see Company 5 also using metaphor of house).

N2: what has happened (in Yanacocha) is a result of the attitude of people, the attitude is very hierarchical, I am the engineer, you are the rural person, right? So then that attitude is based on hierarchy and is obviously linked, you know, it is linked with our history, with our past, racist, viceroyalty with everything that is involved with how the viceroyalty is in Peru, how they gave the, the, how the basis for the Peruvian nation were created, which is a basis based in hierarchy. (In Yanacocha) for them dialogue means I am going to betray you, I am going to come here and talk and I am going to impose this and I am going to mislead you, why?

7.3.3   DISTRUST OF COMPANIES

The people who work for the ONDS, company people who regularly attend community meetings, and the communities themselves all agree that there is a fundamental lack of trust in mining companies. This mistrust is anchored in the past, as described by a senior employee at small mining company:

C6b: When we arrived it was a very conflictive situation. In fact it was, um, basically "get out, we don't want anyone, we don't want foreign investment, we don't want you here." And they had threatened to lynch a person, an exploration geologist and had him in a small hut. They were all sort of deciding as a community, they had him in there, a friend of his from the community went and let him and said "you run, and run as fast as you can and don't come back."

Another area of mistrust lies with the rhetoric that may be employed by company officials. This particular NGO representative is skeptical about the use of words such as "transparency" and "dialogue":

N2: transparency, what does that mean to the company? What is that? So for the company transparency is carrying out an Environmental Impact Study, right? Then that the Environmental Impact Study is presented to the state, then that state approves it as fast as they can, right? Or for example, let's not use the word transparency, but the word dialogue. Yes, so dialogue, what does the state understand by this word? For the state, dialogue means that I have this idea, it is a good idea, this idea produces development, and I am going to have 'dialogue' with you to convince you that this idea is a good idea.

While companies acknowledge there is a history of mistrust, most of those interviewed felt that mining policies and practices had improved since the "old days" of mining but it was difficult to persuade communities about those changes:

C4: look—things are different, things have changed. Now we have really different standards, we are more responsible. We are more formal (I'm talking about mining companies in general) and we really want to show improvement for the whole community around the mining business here. And how we have helped and contributed to that improvement. So convincing families around that long term benefit is challenging because of the past.

7.3.4   COMPANIES DON'T PROVIDE ENOUGH INFORMATION ABOUT MINING BEFOREHAND

According to the head of community relations in a junior exploration company, the lack of transparency around the real impact of mining is a cause of many problems.

C5b: the first thing to mention is that social conflicts begin because of the lack of information. Generally, populations surrounding the mining project don't have good information to have a real perspective about the project itself and its development. So the population has, in previous experiences, bad experiences from the past. If the population hasn't enough information, the first responsibility relies on the company, the mining companies. The population are really threatened because of pollution, because of the dawning of new spaces or new locations because of mining activities. The mining companies, from the very beginning don't send information to the population.

Another company representative felt there was a lack of communication about improvements in technology and different ways of doing mining:

C4: in general terms over the last thirty years technology has improved significantly (such as mitigating dust). Environmental standards have improved significantly. And the way we work has improved significantly. In many companies unfortunately the communications part, the information hasn't improved in the same path. So there is a huge gap between all these improvements from many companies and what the communities perceive. And that gap has been sealed by the lack of information.

The uncertainty and worry that can be caused by lack of information and clear explanation is illustrated below in a quote from a member of women's community-based organization:

W4: our main worry is the water that we drink, we know that in reality it is a water that is not good for human consumption, so what kind of life are we living? with this mining there is too much abuse, to our waters, … the center where they treat the water for the people, however they don't give us information from the tests of the water, the only thing that comes out is colored water, and a boiling Clorox, that looks like milk, that is to disguise the contamination, so, that style of life is what we are living.

7.3.5 HOW COMMUNITIES ARE PERCEIVED

Company perceptions of communities—their culture, their reaction to mining, what they want from mining—are varied because each site is so different, comprising different ethnic groups with different attitudes to mining. As one CEO put it: *each community has its own fingerprint. It's not a recipe for a pisco sour and just mix, mix everything and have a formula for everything. Each community has its own culture, its own problems.*

Despite the fact that communities are the object of very different projects: people to be developed, to be protected, to be defended, for example, there are a couple of broad areas of agreement about the way communities have been treated in the past.

It is also worth explaining the idea of "communities" in the Andes: the people impacted by a mine may not be geographically bounded like villages in Europe, or parts of Asia for example. The inhabitants of such village, apart from living in close proximity, often have a central meeting place: a town square, a church, a town hall for example. This is not the case in the Andes where rural families are dispersed over large areas. As one company representative (6b) put it: *communities are formalized through negotiation with mining companies.*

7.3.6 THEY ARE NOT RESPECTED

Many interviewees—from companies, NGOs, universities and community-based groups—agreed that Peruvian elites (including middle-class engineers) had treated, and continued to treat, rural people in the Andes with arrogance.

> *Company Manager of Community Relations C5b: They feel they are not considered—they are not respected. To first get inside a house, you must first knock on the door. If you do not knock on the door, the owners will feel as if they have been invaded or robbed by thieves. First thing the company has to do is respect the community, the second one is the project. Companies have to disclose all information and give all the information to the community in order to have a real picture of the project.*

Lack of respect is much more than lack of consideration. Many company representatives felt, in common with development ideology everywhere, that education was the key to betterment. Lack of education is seen as a hindrance. Two representatives of one company—that has formed a unique agreement with a community to have a share in company profits—offer a different perspective. They acknowledge that communities have a different world view and are particularly sensitive to slights and disdainful attitudes:

> *C6a: and you know, these people they feel that. What you just said (about being "uneducated") they feel it. If other people feel they are not intelligent, they feel it. They smell it. "What, you think I am stupid? I am not stupid."*

> *C6b: I see various types of intelligence in the world, and lack thereof. And I have had a privileged life, a privileged background but ... there are different ways of seeing the world.*

And in order to be successful when you go in to other communities, you've got to step in and say "where have they been successful." I mean these are difficult situations; difficult histories and they have a different world view. And I have to—I have to—put myself into their shoes and try my best to look from their perspective not the other way round. Just because I come bearing "investment," so what? There's a different way of looking at people, because these are smaller communities, they don't trust us—it's a different way. It's a very different way.

C6b: They have a different sense of time as well. They're not going to run to your company, your western approach, to time, to efficiency. They're not interested. They want to get the sense of you as people, are you trustworthy? "Prove it. Prove it. We don't trust you." We had meetings where people would stand up and say "you white ghosts, what are you doing here? You'll just slip away like all the rest."

Lack of respect is experienced by many rural Andean communities in all aspects of their life. In Chapter 2, we provided material from other sources which substantiated this claim: "Furthermore, *campesinos* (peasants) allege that they are treated with contempt. They have to wait till last to be seen by public officials, they are tricked because they cannot read or write, and they have to show deference for those titled 'doctor,' 'boss,' or 'sir'." (Munoz et al. 2007) [5]. The following excerpt is from an interview with a woman who lives near a mine site in the Cusco District. Her son was injured during a protest in a nearby town in 2010. It is not clear that the protest was necessarily against mining, or even whether her son was an activist, or merely attending the event out of curiosity. Regardless, it conveys the subjective experience of humiliation of trying to get medical treatment in the town hospital:

"in the clinic they discriminated against us because… we are from Espinar, Cusco… The medics didn't wash my son, doing the operation like that, like anything, before putting on the anaesthesia they were already taking the things out, grabbing the pliers and grabbing it, instead of softening it they were…. The doctor said 'put up with it, aren't you Chumvilbicano?, aren't you Yaureno? put up with it, like you put up with your woman,' that is how he said it, that is how the doctor treats us, they are a bit drastic there, 'this is the mother of the rioter, the young man who was shot, this is her,' when I went to borrow a walker so that we could take him out of the clinic, 'please lend it to me, I will return it on the next consultation, please lend it to me, or let me rent it' and then they said we are going to look for someone, and that woman was the one saying 'this is mother of the rioter,' so they treated us like that, right? but why do they treat me like that, no no no, they will never cry like me, they will never suffer, 'I am a humble person, I am from a very sensible family, and, don't treat me like that, it is not my fault, I told them, it is not my son's fault, it was unfortunate that my son had those bullets, don't treat me like that'…"

7.3.7   THEY ARE POOR

Mining companies and government personnel (not just in Peru, but around the world) feel that mining should alleviate rural poverty. According to a Senior Vice President who has worked in Latin America for many years (C2): "based on my life experience of having been born in a mining environment and having lived in it, and worked in it all my life in a dozen countries (I have) seen what a well-considered investment can do to lift people out of the treadmill of poverty." When asked what poverty meant, another company person replied:

C4: the poverty gap is still very high. What you have is 25% of the whole population in the country in poverty but when you go to the rural areas, to the highlands, that goes up to 50%. What is poverty? "they don't meet their canasta basica familia. Canasta basica means—you know what it means right? They don't meet 750 soles a year ok which is their minimum wage. Out of 400 families that are in our operational area, 98 families are vulnerable according to IFC standards. So you do have some people that are considered vulnerable because of the canasta basica, because of the elder's health for example.

Yet there are many cases where the indicators of poverty are worse despite a long history of mining in the area. This type of impoverishment was recognized by most academics, two government representatives, and two company representatives. A selection of their views is presented here:

CEO of Exploration Company C6a: What happened at that stage is that no opportunities were given to people in the community. You have big investments in some areas that probably they were exploiting for twenty, twenty five years. You go back there to those places, to those districts even provinces of Peru, are still poor. After they have extracted over 30 million ounces of gold, how come twenty five years later that city is still as poor as it was twenty five years ago? So actually if you tell me "Diego why a project like Conga, ok, why it was stopped by the citizens?" Because they are paying an invoice. They are paying something they owe to citizens. Twenty five years of non-investment for poor people that need it. And that is the reality of what is going on in Cajamarca. Ok.

Commissioner from ONDS G3: They are poor people. Even though the exploitation and extraction of oil has been going on since 1960, forty years ago, the people always are poor people and the environment is really contaminated so the people they are true, they are right.

G3: In Peru there is a big contradiction. For example, in a little region Bahurubamba we have gas. Camisea project. A big area of gas. But in Bahurubamba live the Machigenga community. Maybe 10 hundred Machigenga people and 10 years ago the malnutrition rate it was I don't know 70% and now ten years later with royalties with cash, with different ways of incomes to the public sector, we have 80%. We don't understand! We have more economic resources but the people are worse. I don't know what has happened. It's a big contradiction.

As mentioned earlier, some companies blame the regional governments for not dispersing the mining royalties for the benefit of the people while others blame specific company policies. In addition to the impoverishment of people impacted by mining, there is a growing gap between those who benefit and those who do not. Two senior company men, one Peruvian (C4) and one American (C2) present their views:

C4: So you have in the community, in the same house you have two brothers—one works for the mining company and the other works on the field in agriculture. And definitely there is a big gap there. And that's a problem sometimes. It's positive if you manage the whole picture and not just with the idea that bringing money to the community will make it positive. So we don't want people to start leaving their crops or leaving their field to work for the mining company.

C2: You have this group of well meaning, well intended gringos who came down there to manage the Yanacocha property and they hired a number of reasonably good Peruvians but they didn't break that ingrained system of "them" and "us." And that permeated through the Cajamarca region, it was poison. All of a sudden these guys are driving around in these nice big cars and the people they used to know are still looking for a combi or a taxi if they can afford it. And stores pop up with these wonderful televisions and things and the only people shopping there are the employees of the mines or their contractors. And that breeds envy. And that's a natural human characteristic and you can't overcome that unless you can give those who do not have that opportunity the hope that your kids can get here if we work together. If we educate your children. But what did they do? They put in a beautiful private school. Davy College. A wonderful school, one of the best in Peru and I have a huge amount of admiration for the quality of education there. But they made it so exclusive that they're helping create—another of Peru's problems—another standard of people. The graduates from Davy certainly are not going to be the same as the graduates of the public schools around there. When it comes time to apply for entrance to the University of Lima, you know who's going to be accepted. Right or wrong, that's reinforcing that pre-existing class system instead of trying to diminish it.

Some people speculated about the possible link between the size of the project and impoverishment.

ONDS Commissioner G4: This is not the best thing but is there a coincidence between the rate of poverty and the big projects? In Cajamarca you have the big rate of poverty and you have projects with a big investment. It is not a happy coincidence but it's a truth. There is a correlation between the big projects with big amounts of investment and big rates of poverty. So this office tries to resolve the gap between the company who wants to get the social license and the population who has to get benefits.

Anthropologist A4: one of my arguments is that the bigger you are the more opposition you are going to encounter, which is totally opposite of how the media portrays it that the bigger you are the more modern you are the more enlightened you are and the more corporate responsibility you have and therefore people are going to get along with you and it is totally not what we see, the big companies which are, have this big expensive corporate social responsibility programs are the ones that encounter the most opposition and that, underground mines much less they have conflicts but they are not conflicts about rejecting mining they are conflicts about more of a cut or getting more jobs and then the artisanal mines they have a big conflict with the state but they don't have so much conflict with the local population because usually the indigenous communities they will work.

Community views about being poor, or being perceived as poor (an important distinction) is that, whatever their perceived status, they should be treated with respect. According to a Commissioner with the ONDS: "Yes. They tell me. The first position always is "we don't need mining, we are poor people but we have dignity. We are poor people but we have dignity. We don't need mining."

7.3.8   THEY HAVE A CLOSE RELATIONSHIP TO THE LAND

Rural people everywhere rely on land and water for their livelihood. The lack of regulation by governments, and the lack of regard for the environment by mining companies (see 3.2.1) has affected rural livelihoods—in some areas severely.

While there may be some who want to work on mines, and have engaged in migrant labor, most farmers want to keep growing crops and raising livestock. Farmers who have experienced contamination of water—which affects crops and livestock—therefore feel that not only their livelihood but their cultural identity is being destroyed.

Farmer from Cusco District F4: principally here, because we live close to the tailing pond, and close to the mine, so, here in the area of X is a red area because of the contamination because there is a lot of water filtrating through …contamination because the land was totally deteriorated, there wasn't like before, produce, like I said before, they declared it an area of poverty, area of poverty in what sense? in the sense that there was contamination because of the tailing pond, it is there next to us, and there is filtrations, not only filtrations but also water is coming out from it, so, before we had sweet water but now no more, it's not only the water but also the air, the dust because the tailing pond is here on our dinner plates, because all of the waste that the company throws away hits here, there are oxides, chemicals, I don't know what chemicals there is still now, because us, I personally don't know.

Community-based Radio Program announcer M6: there isn't going to be cancer to the skin, or climate change, but lung cancer, because they are there exploding so many tons of dynamite, I personally, am against that type of mining I don't think that there can be mining that doesn't contaminate, and if it did exist, it would be so expensive that they would not be interested at all.

A few people perceive a difference in world views, rather than framing the difference as one of educational level or cultural hierarchy:

Lawyer NGO N1: because these people have a special connection with the land and with the water and with the environment. Maybe the people of the city have a special connection with the credit cards and the cars but it's deeper in the rural areas.

Anthropologist A2: there is a real disconnection between the planning of the company and the knowledge and understanding of the community. The company and the community have a different logic and different world views.

7.3.9 THEY ARE SURVIVORS

We have previously referred to the historical lack of respect for Andean people. Only two people, who both work in junior exploration company in the southern Andes which was particularly affected by the Shining Path killings, mentioned this explicitly:

they have had to look after themselves; they have had to survive slavery, colonialism, terrorism. These people they have survived colonialism, slavery and then terrorism. They are still alive and they are still able to produce. How come they are in stress with capital? So are they guilty or are the guys that bring the capital guilty? That's the question. So at the end of the day the communities they are stressed. I think it is the capital that is what stresses the communities. You know why? Because they don't use, what we say, "common sense."

7.3.10 LEARNING TO RESIST, LEARNING TO NEGOTIATE

Even though there is work to be done in how to communicate more effectively about mining, community-based organizations are taking steps to learn more about their rights and laws which regulating mining projects. A woman from a community-based group reflects on her learning process:

W1: during this time in the resistance we have learned some things, we have learned how the state works, or how the mining companies work, for example the mining companies, the communities find out that a company is going to start its work the day that its machines are going into their land, right?

W1: I was taking a diploma about indigenous people and in that I have learned a lot of things, there are processes of granting concessions that the state itself promotes, respecting the 169 agreement, however the mining companies with the state do not fulfil them, for example the state gives a concession and the companies should do this... well they should first indicate the coordinates of the space that the mining companies are going to use, once these coordinates are located, the state has a certain quantity of hectare that it gives to the mining company, and should not go over these hectares, however, we see here in X, that the concessions have gone over those hectares.

Radio is an important means of disseminating information amongst rural people and younger people in community-based organizations are also using Facebook and blogs to share information about the behavior of mining companies. In this case, communities are learning so they can resist. In other instances, the interviewees felt that many communities had become effective negotiators:

Anthropologist A3: they have learned. Right now you have tough negotiators all around the country in Peru. They know from the beginning. Right now for example when one company goes to one town, just to see, or to do some exploration work, they start a demonstration. And if you ask them "do you want a mine" they say "maybe, yes but let us negotiate." So in most cases, it is not a case of they don't want mining. What they want is to get all they can from that mine and that is sometimes a cultural thing. They have learned to take as you go. You can change that also but you need a strong state presence, a change of the rules, and transparency. You need to change the way people—and it's completely different ...the only thing that is common in all the mining negotiations is the way they negotiate, their expectation, is completely different. You go to a small mine maybe here in Lima. They start off they want a truck, they want to negotiate a truck because you know it's a small mine, they don't believe that they have more money and they think "that's all we can get." You go to some other places "well I want one million dollars per hectare" and they learned that. They say "you are not following the World Bank Guidelines so I don't want to talk with you." So that is the level of learning right now in most cases. So you have to know that, to learn that.

The staff at the National Office of Dialogue and Sustainability also encountered different expectations:

G3: Before it was no project, no way, no Conga. In many cases, the first position is confrontation maybe violence. But behind, you have people with expectations about benefits and they wait for the company and the government to share the benefits with them. I think this is the principal change in the minds of these people from five years ago. In the last five years. Maybe before, they would say "no, no, no. The ideologies don't meet in this project because we don't believe in this economic mechanism, no." But now it is "hey share with us your benefits, we need public services. And we have dignity." Very important.

7.3.11   HOW NGOS ARE PERCEIVED

Non-government organizations in Peru have played a mediatory and advisory to community-based organizations and individuals affected by mining. Companies are divided in their opinions about non-government organizations. Some recognize that they fulfill an important role in community life but others believe that NGOs work against peaceful resolution to mining conflicts.

> C2: (talking about open meeting with communities) And I know, I know, I know that there will be NGO groups that are anti-mining, anti-development, anti-whatever—I've never seen one that's pro anything—but I accept that as part of society and I have to be prepared to able to respond in a way that satisfies their question but carries a message to those that have heard the question and understand that these NGOs have been integrated into the community for a long, long time and have talked to the communities for many weeks and months, sometimes for far longer than my people have been able to do, about the evils that will be vested upon them if certain developments take place.

> C4: need to work with NGOs to fill gap about technical knowledge and where companies have made improvements. We have a proactive approach toward NGOs and I am talking about Oxfam, for example, and other NGOs that we have direct contact with. And we like to inform them about what is going on, what we are doing.

One company representative saw some NGOs as obstructive.

> C3: As I told you before in all these conflicts we have had a part, the mining sector, the mining companies have had its responsibility as well. What I don't personally like, as myself, is that there are certain groups—not all ngos because I have worked with ngos before joining the sector and I can make a difference between some and the others—but there are a group of ngos that don't want to engage in dialogue. They don't have the intention of dialoging they don't look for answers, they just want their opinion to be the opinion and their concepts to be the concepts and no other concepts to be put forward. I have worked with Xstrata and I can openly say I had a good relationship with them and we had good, open discussions and I would say a very constructive relationship in the case of Las Bambas (right at the beginning), and to be completely honest I never had any trouble with Cooperaccion even though they were working in the area where we were. That was in the early stages of Las Bambas, in 2004–2007 when we were in exploration, when we were building all the main roads for the early stage of the project. But then also a big part of these demonstrations were discussions more based on perception, than on reality. Then you started having all these different groups and all these different NGOs raising a lot of finances and funds from Europe basically, discussing environmental pollution when in some cases there was no environmental pollution, and changing the discussion in to more of a perception issue.

Another company representative strongly felt that information about mining and mineral processing should come directly from companies themselves:

C5b: If mining companies don't make this effort to show in a way that people understand the meaning of the projects, other institutions like NGOs could be the translators and maybe the information will be deviated from the original. Our company has to be aware that this information is public but this information must be public for the community not just the government.

7.4   LIVED EXPERIENCES OF COMMUNITY MEMBERS

As part of this project we created two short films, based on interviews with women whose lives had been transformed, for the worse, by mining. Sometimes it is difficult when reading a report to understand the perspectives of community members affected by mining. These short films are intended to be used in conjunction with the guidelines and suggestions in this report, to support an understanding of the potential impact of the mining activities on individual lives. They clearly indicate the level of human rights violations which are being enacted on community members in areas of conflict.

The Acuña family has been a highly visible but reluctant symbol of resistance to the expansion of Minera Yanacocha since the latter took the family to court in a 2012 land ownership dispute. After multiple court rulings and appeals, the suit against the Acuñas was finally dismissed in late 2014, vindicating their claims of legal property ownership. In the intervening years and, continuing into 2015 past the final favorable adjudication, the family has been the target of violent harassments, which have included beatings, death threats, poisoning of their animals, the burning of their home, and prohibition against farming their land. Those harassments were orchestrated by Minera Yanacocha and its subsidiaries with the complicity of national police authorities. Without local NGO legal support and international financial and public communications support, Maxima and her family would have remained one amongst many anonymous victims of under-regulated/under-monitored mining activity documented in many of the narratives of this and Chapter 2.

https://vimeo.com/122399156

Melchora Surco Dimache lives in the isolated rural hamlet of Pacpaco, a 40-minute drive from Espinar, the nearest town. She has lived and farmed there all of her life, eking out a subsistence living as did her parents and her parents' parents. Now her property abuts the tailings of the Antapaccay copper mine that, according to Melchora, have poisoned her land, her water, and her animals, making farming and husbandry impossible, destroying her livelihood and way of life. The evidence of independent studies, which tends to support these claims, clearly details the obfuscation and bad faith efforts of Glencore Xstrata's responses. But, lacking the level of support and visibility that aided the Acuña's struggle against a multi-national mining company, Melchora, in her anonymity, has little recourse.

https://vimeo.com/122321445

Although the two interviews differ in detail and locality, the asymmetries of power mirror the similar actor networks: the mining companies in both these cases are global; they have access to capital, to financial markets, and to a legion of international lawyers able to defend their positions; they can call upon extra-judicial policing or special operations units of national police to counter any opposition to their activities; they have the resources to finance research and science, which they present as "objective truth" about the environmental impacts of their operations; they often collude with government to influence regulatory statutes and the enforcement of those statutes and, in worst cases scenarios, are protected by those collusions with immunity from prosecution for human rights violations. Campesinos and indigenous peoples like Melchora and the Acuña family have their land, their air, their water, their crops, their animals, and their social structures, all of which are at risk of being degraded or lost as a result of mining activities; complainants can be criminalized and indigeneity can be legislated out of existence in order to nullify traditional land ownership titles; some people have access to local and international NGOs for representation and legal assistance, but these bodies were marginalized by companies and governments and media in both cases as obstreperous and intransigent "impediments to development."

7.5   WHAT NEEDS TO CHANGE

It is common sense to suppose that suggestions to work toward a more equitable engagement with communities around mine sites would simply require addressing or reversing the negative issues outlined above. The views below are more practical in nature and are based on the experience and mistakes of the people affected by mining.

7.5.1   COMPANY CULTURE

Many of the community members in the previous report complained that companies set up public meetings at which people could not speak, were prevented from speaking if their questions were critical, or where the meetings were stacked with representatives of mine workers, or members from different communities. They acknowledged that "community engagement" workers could be decent people but they often changed jobs and it seemed as if they had no real power in the company. Many NGOs felt that community engagement were the "nice face" of the company which had no real impact on policies.

The idea that companies need to change their internal culture is strongly supported by the academics:

> *Anthropologist A2 (describing his experience as a consultant for a major mining company): Before changes to management, day to day it was a nightmare. There were small conflicts within teams and no communication between teams. With a company takeover that situation changed substantially.*

> *Anthropologist A3: It is a painful learning process. Only a handful of companies have actually changed their internal culture. Most of them it is tokenistic ... So I would say that*

just a handful of companies in Peru have jumped up, have upgraded … 5 companies, 7 companies … no more than that. The rest of the companies, yeah, it's business as usual or sometimes they hire one guy or they hire an anthropologist, they try to improve a little bit but it's not like real approach. It's just something to make up.

The following suggestions for internal change come from companies who, although they have encountered conflict when they acquired their lease, are at a point where they are not encountering social protest on the level experienced by communities in the previous report. These do not follow global guidelines (such as the ICMM) but reflect the institutional culture of each company.

7.5.2   INVEST TIME NOT MONEY

Company 4, held up by many as an exemplary case of how to go about establishing relations with a community has spent six years in exploration and negotiating with local people. Company 6 is also still in exploration and has completed an Agreement with the communities to be shareholders in the company.

C6a: And these were long winded discussions that were held outside of the community to start with because we weren't welcome inside the community. As trust started to develop from those discussions, we were started to be invited in and to talk to the wider group of the community. But what we discovered was that the people holding the power were not giving any benefits to these groups and we realized "ok here's a point that we can work with." So we then basically decided to—and it took—because we didn't get to the 5% right in the beginning, this is something that developed. We had a very good community rights lawyer who came with us, who had worked in conflicted areas. As we started to draft up agreements, sometimes we'd use the wrong words and there would be conflict again. He would then—he would look at other words because it had to be legally binding for both parties but also in language they could understand. It took a long time for us to get to know them and for them to get to know us. And then you suggested (referring to the CEO, also at the interview) that they need to, as a community, look at the benefits they could gain. Not just a few miners but the widespread community that could then start to get benefits, have surface rights payments.

The CEO of Company 2 travels to the mine site fortnightly, if not weekly:

C2: money alone is not the answer. If people don't have faith in you they're not going to have faith in what you do with your money. And so for previous mines where I worked in Peru, I personally invested hundreds and hundreds of hours in the communities. Me, along with my team, developing a trust base and as the head of company they began to understand me, they began to trust me—not always agree, I don't always look for agreement but always respect me. Respect me. And then that trust and respect is transferred to the people that are working with me. My people have to be part of the communities—not just on a physical level which I insist on—but through being an integral part of community affairs. You have to take a broad view of what you're doing. Why did Company 1 not have a problem, why is Company 2 seen as good? We spend 1% a year in terms of dollars of what Company 4 does but we spend 500% more time in the communities and with the people than they do.

7.5.3   SENIOR PEOPLE MUST MEET COMMUNITIES FACE TO FACE

C2: I've spent 11 hours in a community meeting for example with hundreds of people and those are difficult. Not everybody's my buddy. There are people who want to hit me and do other things. I ask them to meet me—300, 400 that's fine, let's go out into the field and have a meeting. I'm here to listen. Sometimes that's all I do. I say "I want to listen to what your problems are, I'm not bringing a solution, and let me tell you up front I'm not going to give you anything but I want to understand what the issues so maybe we can work together to resolve some of the problems." So I go to the meetings. And they're very often not something the local elected authorities want to happen because they see it—and they're correct—as reducing their individual authority. If they're not the ones carrying the message, what authority do they have? But if I can't depend on them to carry the message correctly, how can I use them? … (and) I tell them in the meetings as I point to my head of community relations or someone else, I say "you know when he talks, those are my words." And that puts a huge responsibility on that individual too. But it also shows that people with whom I am dealing, with whom I've developed a level of trust and understanding, that I have faith in this person. And so when he makes a promise, it's the same as if I had made that promise.

C6b: And so we (6a, myself, a driver and a few others) arrived and there was this very conflictive situation. So and we began slowly through many months of negotiations,—we went in ourselves. So we didn't send in a little team of people that couldn't make decisions. That couldn't actually say "this is …" or that would make a decision and then go back to their board and the board would say "no we can't let you give them that" and then you'd have that whole issue of lack of trust again. So we went in, we went in regularly, we would have six hour long meetings and then we started to work out how this community was operating. They are very, very long winded meetings. Very long meetings. You have to put the time in and for them it's a little bit like theatre. You become the entertainment

C6a: Ok we have to be fair. All this, in house people that works the community relations, in big companies are outsourced, they are hired to fix things. These guys they receive orders from a board of directors sitting in Toronto, Vancouver, London. People have no knowledge but only knowledge of budgets and timing. Pick up the phone, tell a guy "you know I need this in fifteen days" and the poor guy, probably an engineer a gringo that works for the company here, has to go the community makes such a pressure because he doesn't have the time.

7.5.4   SHARED VALUES WITHIN A COMPANY

Three mining companies also stressed that attitudes toward the community—treating them with respect, and respecting their values—must permeate the whole company.

C2: It's the selection of the right people. I'm not going to select everyone, that's foolish. But if I have a core group who understand and believe in what I think is necessary, then I've accomplished what I really need to because that gives them the responsibility that the people they select to work with them also either understand and believe or are capable of learning the practices. And that's a real challenge. The people that I have working for me now, I've had some that have worked with me in West Africa, some that have worked with me in Venezuela … They've come with me and I think it's because they believe in the style of development that I've used.

C5b: They feel they are not considered—they are not respected. To first get inside a house, you must first knock on the door. If you do not knock on the door, the owners will feel as if they have been invaded or robbed by thieves. First thing the company has to do is respect the community, the second one is the project. Companies have to disclose all information and give all the information to the community in order to have a real picture of the project.

C6b: you can bring roads and infrastructure but this is not about building a relationship. The other company, (before them) the big roads I mean the logistics of the place is amazing— instead of bringing the best out of the community, prostitution was starting up with the foreign workers building this road. And that's the thing. You've got to have very strong leadership that can really—if you have a sexual harassment case, you've got to kick it as hard as you can and get it out. If you have someone mess around with the community, you've got to stamp down on it. And because we were there and because we were able to make decisions and keep promises and then trust starts to get developed. And it takes a long time.

While the idea of "shared values" sounds like empty rhetoric it only has the possibility of becoming a reality if senior staff are present in Lima, and are able to make regular trips to the proposed or existing mine site, towns or communities. All the open meetings described in the interviews were held in open fields, until (for Company 6) they were held in a town hall, the first project requested by the local people.

7.6   COMPANY POLICY

7.6.1   COMMUNICATION ABOUT THE IMPACT OF MINING

Company 4 has taken community groups to visit other open pit mine sites to see the scale of mining operations. Other companies have not done that but are also concerned about communicating the scale, noise, dust, etc. that is associated with mining.

> *C2: (speaking about qualities in the CE team) Far more than sociology or psychology or some of the others because it's vital that the person have the ability to communicate the concepts of the changes that are going to take place in a mining environment. Because so many of these ore deposits are found in remote places that are so quiet at night you can hear a kitten walking. And they're so remote that electricity is years away were it not for the coming of the mining company. And he needs to be able to explain engineering concepts but in plain language what these changes are going to be. If we just say "oh it's going to be wonderful and beautiful, you're going to have electricity, etcetera" but doesn't tell them "you're going to hear trucks all night long," then that's not all the story. And pretty soon the adversaries, those that are well-meaning and those that are malicious, will start pointing out "come over to this mine at 10.00 o'clock at night and tell me if you can sleep with that noise in the background." Or "look at this huge pile of rock waste that's here, how's that going to look in your community?" So you have to be able to explain what are the long term ideas for remediation when the mines go away. They all go away and they all leave a large and visible impact.*

> *Company 5a (describing the way they engage with communities): we are making some presentations to the community and in these presentations we show by pictures some kind of the drillings and the equipment but also we arrange some visits to the field to show people how the drills operate and what kind of impact it has. So there is a chance for villagers to take note in the field what it looks like, the mining activity. Even we have some people who were reluctant to the project but when they have the chance to look at the rigs, and the drilling activities, they could feel and make sense of the speech they had heard before.*

> *Anthropologist A2: We also need to explain about mining in schools. In all the regions where is mining, they explain nothing about mining in schools in the general education. If mining is going to in your region for many years you need to understand how mining works. For primary children for secondary children, so they can explain this to their parents. But no it is absolutely disconnected.*

7.6.2   COMMUNITY MONITORING OF ENVIRONMENTAL IMPACT

The issue of water contamination is a highly contentious issue for many communities, particularly in the District of Cusco. Only one company had commenced an environmental monitoring program in collaboration with the *rondas* and farming families, and this program was implemented with the assistance of an anthropologist (A1). This is the same company that has spent 7 years in the exploration phase. Their Integrated Water Planning Management Group includes: the *rondas*, the company, a community-based organization, and a local technical education institute.

Another anthropologist was cautious about the value of environmental monitoring programs:

> *A4: It is making the issue very technical and that is not going to help explain the conflicts. I am all for determining through scientific means whether there is contamination or not, I think that knowledge would be very useful for communities and to the country as a whole but that's not usually what explains whether there are protests or not people now protest because they objectively know that the river is being contaminated they may perceive that but whether it actually is or not is a different matter.*

7.6.3   SOCIAL IMPERATIVES SHOULD OUTWEIGH FINANCIAL IMPERATIVES

Many company people said that social issues were equally as important as technical issues.

> *C2: Let me put it in the context of … if I have an engineering problem at my mine, it's an embarrassment, it may cost me millions of dollars to fix. If I have a major social problem, I might lose the mine. I might lose the company! I darn well better make sure that what I'm doing is correct.*

Yet—and this is a crucial issue which was raised in Chapter 2—would social issues "trump" financial imperatives? It has always been major stumbling block (as seen in the Ok Tedi case in Papua New Guinea) for companies to forego profits in order to deal with complaints about either social or environmental issues. Academic 3 feels that only a handful of companies currently operating in Peru would put aside financial imperatives.

> *A3: still, social issues are not the main issues. Even if they say "no it's our main issue" because we have had some conflict but—when you know you have to make a hard distinction you look at the figures, you look at what is business and what is not, you look at timing, you look at the operations side, at the engineering side, but still you look at the social side in the same way. For example, "we have to decide whether we are going to build up let's say a new open pit right now and we have to delay this decision for the future" and they ask the financial people "do you believe we are going to get profit right now? Or if we do that afterward are we going to lose money?" And they look at the engineers, the operations side "what do you say?" And afterward they say "ok social side you have to accommodate with the others decision" and that is a huge handicap.*
>
> *In that sense, what happens is that you start the operation and then you have to stop the operation. You have to stop the operation because of a huge conflict and you cannot deal*

with that. So the only company I think that has had some talk about that is Rio Tinto. So they say "it's not socially possible right now so we have to delay our decision even if we are going to get profit, even if we have all the machines ready—there is a social concern that we cannot surpass." They started the new kind of relations. So this is also why this case is so different from Yanacocha for example. Both of them are in the rondas environment, they are in Cajamarca—they have the same thing. But they are completely different. Because you have one place in which companies started to change, and change for good, and the other place that didn't happen.

7.6.4 STOP MINING

There were only two company representatives who believed that, were all of the above to fail, that mining should cease.

C2: Because not every mine in the world, or should I say every mineral deposit in the world, should be a mine. There are some that should be excluded for now. Ultimately the needs of man will prevail and they will be exploited that's almost a foregone conclusion, but we have to take the moment into consideration as well.

This view is supported by farmers and community-based organizations:

M5: it is not that we are really against mining here, I don't want them to get that impression, no, but the problem is that the place where they want to do the extraction is unviable, so… it is where the water is born for everyone, and for them to do that there it is practically leaving us without anything, right? because it is obvious that the water will dry up completely, that is the reason, it is not because we don't like mining or that we are angry, like they say… we don't have that anger, rage, right? the problem is that it is unviable the place.

If conflict arises after exploration or mineral extraction has started, and all measures have failed, then companies also agree that mining should cease:

C4: if we don't, you know, get the, let's say the social consensus around working together to develop a mine, (this company) won't develop it. Ok. As simple as that. Because we cannot be developing a project that is stopping every month, or stopping every six months because the community didn't understand this or they didn't understand that or with politician raising common denominators with no basis, you know. That's the thing. So we are taking our time. We're making sure that we're very clear around what we are doing, very clear around the activities, very clear around how we are doing it.

C4: we want to do it properly, we are not perfect, but if we see there are too many restrictions, from a political point of view, (this) is a company that will say "let's leave it there, let's go somewhere else." So that is why we are taking so long. Because we are interested in this project, in this country, and we want to make sure that the technical and social risks are well managed.

Many communities who been negatively impacted by mining for some time support the idea that mining should cease. This is the view expressed by a member of a community-based organization in an area with a long history of conflict, and violent encounters with police:

M1: it seems that generically, issues can be mixed up, but here we don't want jobs, we want them to respect us, we don't want them to say hello to us, we want them to go away from here, and we want clean water like we have always had, so from this point of view there is nothing that can be talked about, there is no point to start the dialogue nor an agreement for them to break, we are conscious that this issue has many axis, right? including ecological technology, social, political, communication, etcetera, but we have said from a long time that Celendin has a second degree conflict, it is not about more or less money, or the possibility of jobs in the mine, also that issue about the verbal agreements, those contracts, they have never met their verbal contracts, they haven't even met what has been written, and that is why we maintain, well at least I do, reject all plans that try to convince us or change our minds from the decision we have made.

7.7 THERE SHOULD BE INDEPENDENT MEDIATION

Academic A believes independent mediation is necessary:

A1: we see mediation as an activity that needs to be mainstreamed in public policy. And in the area processes that may lead to improve the use of natural resources and of course the quality of life for the people. We argue that in Peru this is part of the human nature, and part of our social relationships so it needs to be understood from the very early stages of the project. It needs to be inside and articulated into the decision making process, and in human relationships, for policy purposes. At that stage of thinking, we believe that even if you have a council of mediators, that the process will always be case-based. We need to address the differences between people. It needs to be part of the governance system at every level. Our recommendation is that the Min of Env develop a program with all the stakeholders that participate in the by law process, in each region, to introduce the concept of conflict, reparation time, and responsiveness through independent reviews.

There was little discussion of this idea among the company representatives. The government personnel supported the idea of mediation but believe that the ONDS is the appropriate body to mediate a non-violent resolution to conflict.

G3: and the key element in this period of time was the violence, too much violence. The scale of the conflict, too much violence, and the government in this point well we had to do something about this conflict when the social conflict was increasing. Now we work in this stage when the problem is a big problem, prevention is our focus.

7.8   STATE POLICIES AND VALUES

7.8.1   THE STATE MUST BE INVOLVED FROM THE BEGINNING

This is an area of agreement between the communities and companies that we interviewed.

> *Member of community-based group W1: the state should come to the place where the concession is going to be, so that they can see if it is valid to give the concession in that area or not, because a concession is given blindly, like I have said, they don't care about our identity, because there are some archaeological sites, and within the law it says that a concession cannot be given in the archaeological sites, however they don't care, they don't care about the number of families that live there, what do they live from? what is their way of life? how do they survive? what do they cultivate? so, they give this area as a concession, but the state doesn't know what is part of it, so that is what I think would be a way for the state itself to intervene and in some way, to see what it is given in the concession.*

7.8.2   THE STATE MUST PROTECT ITS CITIZENS

Companies and communities have different ideas about the role of the state vis a vis the communities. Companies feel that the state (particularly regional governments) have neglected community development and misused mining royalties, that they have not "looked after" the interests of the people. Communities who oppose mining feel they have not been protected from the police, or that their complaints have not been taken seriously.

7.9   COMMUNITIES WOULD LIKE TO BE EDUCATED ABOUT MINING

There is a clear desire for communities to learn more about mining, both the regulatory framework as well as the stages of mining, and expectations about impact. One of the interviewees runs a community-based radio program in Quechua language. She expressed an interest in hosting a program about mining for the local communities.

7.9.1   EDUCATION IN SCHOOLS

The companies we interviewed were in agreement that there needs to be clear communication about the real scale and impact of mining. This was also articulated by community people in both the first and second round of research. There is also a clear desire to learn more about mining, both the regulatory framework as well as mineral processing and what will happen with mine closure.

A Peruvian anthropologist and consultant who has looked at the impact of mining on rural areas, felt that education about mining should be included in primary and secondary curriculum:

> *A2: We also need to explain about mining in schools. In all the regions where is mining, they explain nothing about mining in schools in the general education. If mining is going to in*

your region for many years you need to understand how mining works. For primary children for secondary children, so they can explain this to their parents. But no it is absolutely disconnected.

7.9.2   ENVIRONMENTAL MONITORING PROGRAMS

Although it has a narrow focus, there is also opportunity for communities to learn about the impact of mining on water sources; as indicated earlier Company 4 has instigated a collaborative program whereby communities monitor the water in their region.

The material which we have presented in this, and in Chapter 2, presents clear areas of agreement about what needs to change. The statements in the preceding sections provide an insight into what can occur when companies do not take community engagement seriously nor make changes to their behavior and policies. The table above summarizes the guiding principles identified during the first stage of the project and the suggestions for action as identified in the second stage. The key message is that community engagement takes time, that companies must be honest about the potential and real impact on the environment after independent evaluations and they must take responsibility for damage caused,. Furthermore they must respect ways of living that are not the same as their own. This can only be done if senior company executives are committed to change, and that they develop strategies to build shared values within the company from the CEO in Head Office to the local Mine Manager and security staff.

Ultimately it is this action over many years which will build community trust that their views are being heard and respected. Governments, both local and regional also need to be involved and to be seen as first and foremost protecting their citizens, not company interests.

Of course it is true that the number of people who will benefit from the taxes generated by natural resource extraction will far exceed the number of indigenous peoples living in voluntary isolation who will be adversely affected. It is also very likely to exceed the number of indigenous and other people whose livelihoods and territorial claims might be compromised. The question that is not asked, however, is whether one person's right of access to education and health services is equivalent to another person's right to existence, to an ethnic group's collective right to territory or, even, to nature's right to existence. Posing this question does not mean that one automatically concludes that these latter rights (to existence, territory, identity) necessarily trump rights of access to social services. It does mean that this question about how to discuss, agree upon and legitimize trade-offs among these different sorts of rights has to be addressed head on and debated seriously in the public sphere (Bebbington, 2014).

None of the above is straight forward or without cost. But human rights violations simply cannot continue in the guise of development of a nation. This chapter has highlighted some of the important steps which companies and Governments might take if they are to engage more equitably with communities, and to treat them with the respect they deserve. How to help engineering companies and Governments as well as engineering students realize that this is

Table 7.1: Guiding principles (*Continues.*)

Community and NGO Views: Stage 1	Company and Government Views: Stage 2
Transparency	
The need for transparency refers to the need for a clear understanding of the kind of work that is going to be carried out, how long it will take and what impact it will have on people and environment.	Companies must communicate honestly about the time frame and real impact of mining in terms of scale, dust, noise etc. Companies should collaborate with communities in environmental monitoring (5.2.2).
Respect	
To be treated with dignity not as an obstacle to mining.	Companies recognize that communities have not been, and are not, treated with respect and this must change; values of respect must be shared within a company, not just at the executive level. The state must protect its citizens.
Respect for local opinion about where to mine or not mine.	A company should recognize when it must stop mining.
Respect for views about alternative forms of development or a "good life."	
Dialogue	
Interaction with the "company" comes in a variety of guises: with engineers, with managers, with police, with the Public Prosecution Office, and sometimes with community engagement people. Not only do people want a chance to speak directly to those who make decisions that affect their lives, but they need to know who those people are.	Senior company executives must meet communities face to face in open meetings. The state must be involved from the beginning.
Effective dialogue should encompass every section of the community.	Senior company executives must meet communities face to face in open meetings.
Effective dialogue requires a review of time frames.	Companies must invest time not money in community relations.

Table 7.1: (*Continued.*) Guiding principles

Trust	
There are many references to broken promises in the narratives which generate a lack of trust and a general skepticism about the honesty of mining company employees; the integrity of mining company employees; the ability of the state to protect community interests.	All of the above will build trust over time if done well.

optimal way to engage is not always obvious. We hope that this book assists you in this endeavor. In Appendix B we have also given links to short films which we have created about some of those we interviewed, which can be downloaded free of charge to bring some of the people in this book and their stories to life. We have also given links to an interactive game which we have created, based on a number of real life mining community contexts which can also be used to help participants begin to understand what it feels like to be a community member in the midst of the first ten years of a mining intervention.

7.10   NEWS ARTICLES AND OTHER PUBLICATIONS

[1] Boesten, J. (2012). The state and violence against women in Peru: intersecting inequalities and patriarchal rule. *Social Politics: International Studies in Gender, State and Society*, 19(3):361–382. DOI: 10.1093/sp/jxs011. 184

[2] Fischer, M. (2003). Emergent Forms of Life and the Anthropological Voice. Duke University Press.

[3] ICTJ International Centre for Transitional Justice (n.d.). *Peru's Painful Mirror.* https://www.ictj.org/perus-painful-mirror/ 184

[4] Li, F. (2009). Documenting Accountability: Environmental Impact Assessment in a Peruvian Mining Project. PoLAR, 32(2):218–236. DOI: 10.1111/j.1555-2934.2009.01042.x. 179, 180

[5] Munoz, I., Paredes M., and Thorp, R. (2007). Group inequalities and the nature and power of collective action: case studies from Peru. *World Development*, 35(11):1929–1946. DOI: 10.1016/j.worlddev.2007.01.002. 188

APPENDIX A

Mining Companies

Eight mining company representatives who worked in the area of community engagement were interviewed in Lima. Four of these companies operated in the same region in the northern Andes; two operated in other parts of the Andes; and the remaining two worked for the same company operating in southern lowland Peru.

The purpose of the interviews was to ascertain:

- the level of social conflict, if any, which existed at their mine sites;

- the kinds of explanations put forward for the existence/lack of social conflict in Peru generally, and on their mine sites in particular;

- company perceptions of the government's role in managing mining and managing social conflict; and

- the practical steps which different companies take toward engaging communities and resolving conflicts that arise.

The following section sets out a summary profile of each mining company along with the codes for individual/s interviewed from each company.

Companies 1, 2, and 3 operate in the same region. Company 1 has experienced minor conflict (in the form of initial opposition to company takeover); Company 2 has not experienced conflict while Company 3 has experienced extensive conflict. Company 4 is a major mining company, still in the exploratory phase in an area to the north; it also experienced initial conflict but has been involved in community engagement for 7 years while exploring the lease, and there have been no major conflicts during that time; if the project goes ahead, it will be a major project.

Company 5 is in exploration in the southern lowlands of Peru while Companies 6 and 7 operate in the southern part of the Andes.

A.1 COMPANY 1

This is an established mid-tier company operating a copper/gold deposit. There is one open pit, and one copper-gold plant at elevations ranging from approximately 3,600–4,000 metres above sea level. Stage: extraction. The mine site is 1.5 km from a small town of approximately 2000 people; this town, together with 4 rural communities, are designated as being impacted by the mine. This area has a long history of mining, unlike Company 3; there is a history of interest in, and wanting to benefit from, mining activity.

C1 commenced operations 10 years ago at a time when there was significant opposition to mining at another nearby mine site. The town and the rural communities were not initially accepting of Company 1: a presidential election was taking place together with increased social activism against mining across Peru. In 2004, "the communities say 'we don't want new mining because the old mining that worked here was very bad' … and for this reason when our company arrive in 2004 the communities say 'first the new mining need to make remediation for the whole environmental liability'." This remediation is being carried out in partnership with the national government. In August 2013, the sixth amendment to the Environmental Impact Assessment (EIA) concerning extensions to the pit, a new quarry and one topsoil dump, was approved. The projected closure date is 2023.

Interview codes:

C1a: Vice President, Corporate Affairs C1b: Manager of Sustainable Development.

C1c: Manager of Community Relations on mine site.

A.2   COMPANY 2

Mid-tier Chinese joint-venture company. Ore deposit: copper/gold, predominantly copper. Extraction has commenced although EIS approvals are in submission to mine other leases on the site. Projected life of mine: 22 years.

Interview code:

C2 (Senior Vice-President, General Manager).

A.3   COMPANY 3

Major mining company with sites around the world; history of escalating social conflict, including a state of emergency in the main town. Plans to expand into an adjacent region has met with opposition from local communities in the area.

Interview code:

C3 (Newly appointed Director of External Affairs—4 months in position).

A.4   COMPANY 4

Major mining company in exploration phase for 7 years, situated to the north of Companies 1–3 in the northern Andes (altitude 2,455 m). It is the third mining company to operate the lease even though extraction has not yet commenced.

The project began as a state-owned operation, then was owned by a mid-tier Canadian company. Community feelings toward the Canadian company were extremely hostile at this point. A transnational major began re-negotiations before selling to Company 4. There are 44 communities which would be impacted by the proposed project. Land is owned individually, unlike communities impacted by Companies 1–3; the company is leasing land from 31 families to carry out exploration. Some families have opted to resettle to the coast. It will be the first mining project in the area.

Interview code:

> C4 (Newly appointed Director of Corporate and Social Affairs—in position 6 months).

A.5   COMPANY 5

Junior Exploration Company. Extraction has not commenced. Low population density in southern Peru in semi-arid conditions. Initial hostility when Company 5 took over the project.

Interview codes:

> C5a: CEO.

> C5b: Manager of Community Relations.

Both have worked together previously on a mine site marked by extensive social conflict.

A.6   COMPANY 6

Junior Exploration Company operating a gold deposit in the southern Andes (2,785 m). The company faced intense community hostility after buying the lease from a major mining company in 2006. This is the only Company in Peru to have formed an Agreement with the community in which the community are shareholders in the company. The Environmental and Social Impact Assessment (ESIA) was approved in September 2013. The Construction Permit, the final significant permit required to build the mine, was granted in June 2014.

Interview codes:

- 6a: CEO.

- 6b: Former Head of Community Relations.

A.7   COMPANY 7

Company 7 is large mid-tier joint venture mining company. It completed the acquisition of this large scale copper project from a major finance company in mid 2014. The development of the mine site required extensive resettlement, involving 500 families (approx. 2000 people).

Consultation, under the previous owners, started in 2009, an Agreement was reached in 2010. The process of resettlement started about a month before these interviews took place. They maintained the CE model which had been developed by the previous company and maintained the same staff. The resettlement was carried out by a private consultancy company. Before 2009 the community didn't want the project.

Interview code:

C7 (Director of Operations, with previous experience in Australia, India, and the Philippines).

A.8   ACADEMICS AND CONSULTANTS

Four academics, three of whom worked as consultants to mining companies, were interviewed in Lima. The purpose of the interview was to ascertain

- their overall view of mining and community engagement in Peru;

- their particular experience of mining and community engagement; and

- their explanations for the rise in social conflict in mine sites.

The Interview codes are as follows.

- A1: has a Ph.D. in Anthropology and is a private consultant specializing in mediation of social conflict on mine sites (e.g., the Tintaya Roundtable). He has a small team working for him.

- A2: is an Anthropologist with a Master of Social Management; he has 12 years' experience as a consultant to major mining companies working in Peru. He operates his own consultancy agency.

- A3: has a Ph.D. in Anthropology, works at a University in Lima and also for an NGO development consultancy group.

- A4: has a Ph.D. in Anthropology, teaches in a University in Lima, has done research on mining and labor in central Peru.

A.9   GOVERNMENT REPRESENTATIVES

Four government representatives were interviewed including two from the Ministry of Energy and Mines, and two from the National Office for Dialogue and Sustainability (NODS). The NODS was created only two years ago. Its goal is to prevent, and resolve, social conflict on mine sites. According to G3 (see below) they dealt with more than 100 cases and they employ up to 50 commissioners.

The purpose of these interviews was to

- ascertain their role in the regulation of mining and/or the resolution of social conflict and

- their personal views on the effectiveness of their role and the obstacles and challenges which they face.

The interview codes are as follows.

- G1: former member of Social Management Committee, Ministry of Energy and Mines. The Ombudsman's Report was an important source for this committee, a guide to existing and emerging social conflicts.

- G2: Office of Environmental Regulation, Ministry of Energy and Mines (lawyer).

- G3: Commissioner for the National Office of Dialogue and Sustainability (law background).

- G4: Commissioner for the National Office of Dialogue and Sustainability—NODS (social psychology background).

A.10 COMMUNITY-BASED ORGANIZATIONS

Interviews were conducted with six representatives of women's organization in the Cajamarca district, with four representatives of an activist organization that wishes mining to cease, and with three individuals representing different community organizations who are also critical of mining.

The interview codes are as follows.

- W1, W2, W3, W4, W5, W6: representatives of community-based women's organization against mining and in defense of human rights. The group is local but has some international contacts through activists in the UK, France, and Italy.

- M1, M3, M5, M7: representatives of community-based organization which also runs a radio program to inform local people about the social protest against a particular mine in the region.

- M4: young mother representing community-based organization of young people in a small rural town.

- M6: blogger, student activist.

- M7: representative of local organization whose aim is to provide legal advice to people affected by mining, and document breaches of human rights violations and environmental degradation.

A.11 FARMERS

Four farmers were interviewed from the Cusco district and one from the Cajamarca region.

- F1: farmer, Cusco district. She lived next to the tailings dam of the mine site.

- F2 and F3: husband and wife farmers, Cusco district whose son was injured in a protest against the mine.

- F4: farmer, Cusco district.

- F5: farmer, Celendin district.

A.12   NON-GOVERNMENTAL ORGANIZATIONS

- N1: lawyer, NGO.

- N2: lawyer, NGO.

APPENDIX B

Educational Tools for Change Created by Our Team

Two films intended for use as pedagogical tools for engineering students, government personnel, and mining companies have been made by filmmaker Eric Feinblatt. They are embedded in the text above and are also listed here:

https://vimeo.com/122399156
https://vimeo.com/122321445

A Mining and Communities Game has also been created (Armstrong, Baillie, and McKenzie 2014), a description of which is available at:

http://im4dc.org/wp-content/uploads/2014/09/Baillie-Armstrong-game-FR2-Completed-Reportpdf.pdf

Authors' Biographies

CAROLINE BAILLIE

Caroline Baillie is Professor of Praxis in Engineering and Social justice at the University of San Diego, and co-founding director of the not-for-profit "Waste for Life" (wasteforlife.org) which supports vulnerable communities in the development of upcycled waste-based businesses. Baillie's research considers socio-technical processes and systems, which enhance social and environmental justice, and educational systems that promote these. She brings lessons learned from these studies and practices into the classroom of all ages, to facilitate the transformation to a more equitable and just future. Professor Baillie has published 27 scholarly books, over 200 book chapters, peer reviewed journal and conference papers, and is Editor of this book series *Engineers, Technology, and Society*.

ERIC FEINBLATT

Eric Feinblatt was the visual and audio documentary investigator of the IM4DC funded research project in Peru and the Forum for Mining and social justice held meeting at Casa Pueblo in Puerto Rico discussed in this publication. Two of his video interviews—with Maxima Chaupe and family in Cajamarca and Melchora Suco Dimache in Espinar—are described and appear as links in this book. His work about the community impacts of mining were used as evidence in support of these human rights defenders. Mr. Feinblatt is co-founder, along with Caroline Baillie, and Director of Waste for Life, a U.S.-based non-for-profit organization.

JOEL ALEJANDRO MEJIA

Joel Alejandro Mejia is an assistant professor in the Department of Integrated Engineering at the University of San Diego. He worked as a metallurgical and project engineer for Rio Tinto and FLSmidth on different projects around the world, including mining operations in Peru, Mexico, and Zambia. His current research has contributed to the integration of critical theoretical frameworks and Chicano Cultural Studies to investigate and analyze existing deficit models in engineering education, and the development of critical consciousness among engineers. Dr. Mejia's work, influenced by his previous work as a metallurgical engineer, also examines how asset-based models impact the validation and recognition of students and communities of color as holders and creators of knowledge. His work seeks to analyze and describe the tensions, contradictions, and cultural collisions many Latinx students' experience in engineering through *testimonios*. He is particularly interested in approaches that contribute to a more expansive understanding of engineering in sociocultural contexts, the impact of critical consciousness in engineering practice, and development and implementation of culturally responsive pedagogies in engineering education.

GLEVYS RONDON

Glevys Rondon worked for over two decades as director and founder at the Latin American American Mining Monitoring Programme (LAMMP), a London-based organization aiming to empower and improve the lives of women affected by mining in rural Latin America. She is an independent researcher and consultant with expertise in rights, gender, and corporate accountability, particularly in relation to land and natural resources.

JORDAN AITKEN

Jordan Aitken graduated from the University of Western Australia in 2016 with Bachelors' degrees in law and mining engineering. During the course of his studies, Jordan discovered a passion for international law. It was the confluence of international law and mining operations—in particular, their impact on surrounding communities—that formed the basis of his chapter in this book, which was authored during his final year of study. Since graduating, Jordan has worked in several roles across government, including in foreign affairs as an international lawyer, and more recently as an adviser to the Minister Assisting the Minister for Trade and Investment.

RITA ARMSTRONG

Rita Armstrong is an anthropologist whose postgraduate research was based on fieldwork in Central Borneo. She has since worked as a researcher on an "Engineering Education for Social and Environmental Justice" project with Caroline Baillie, and has also taught across the disciplines of Engineering, Anthropology, and Geography around the issues of environment, conservation, gender, and development.

VICKI BILRO

Vicki Bilro completed a Bachelor of Environmental Engineering and then a Master's in Civil Engineering where her thesis topic investigated the impact of mining on rural communities in Peru. Following her studies, she is now working as a Site Engineer for a commercial construction company in Perth.

ANDY FOURIE

Andy Fourie is a Professor of Civil & Mining Engineering at the University of Western Australia. His research focus is the minimization of impact from mine waste and municipal solid waste storage facilities. Prof Fourie's work has contributed to an improved understanding of the issues contributing to catastrophic failures of tailings storage facilities and he has developed techniques to minimize and mitigate these issues. Through the application of geotechnical engineering principles, he has demonstrated how significant dewatering of mine tailings cannot only save water, but contribute to improved stability and potential rehabilitation of these waste landforms. His work on closure of tailings storage facilities, together with colleagues from plant biology and soil science, has demonstrated the potential successes achievable through a multidisciplinary approach to mine closure.

KYLIE MACPHERSON

Kylie Macpherson holds a Master's degree in Mining Engineering from the University of Western Australia, where she pursued her interest in the social and economic impacts of mining. Growing up in a rural Australian town, Kylie was exposed to both indigenous culture and the ways in which industry can impact the outcomes of many—often not equally. Kylie currently works for BHP in a fly-in fly-out role where she is learning about life in the Pilbara Region and mining operations.

Printed in the United States
by Baker & Taylor Publisher Services